International Political Economy

Series Editor: **Timothy M. Shaw**, Visiting Professor, University of Massachusetts Boston, USA, and Emeritus Professor, University of London, UK

The global political economy is in flux as a series of cumulative crises impacts its organisation and governance. The *International Political Economy* series has tracked its development in both analysis and structure over the last three decades. It has always had a concentration on the global South. Now the South increasingly challenges the North as the centre of development, also reflected in a growing number of submissions and publications on indebted Eurozone economies in Southern Europe.

An indispensable resource for scholars and researchers, the series examines a variety of capitalisms and connections by focusing on emerging economies, companies and sectors, debates and policies. It informs diverse policy communities as the established trans-Atlantic North declines and 'the rest', especially the BRICS, rise.

Titles include:

Caroline Kuzemko
THE ENERGY SECURITY–CLIMATE NEXUS

Hans Löfgren and Owain David Williams (*editors*)
THE NEW POLITICAL ECONOMY OF DRUGS
Production, Innovation and TRIPS in the Global South

Timothy Cadman (*editor*)
CLIMATE CHANGE AND GLOBAL POLICY REGIMES
Towards Institutional Legitimacy

Mark Hudson, Ian Hudson and Mara Fridell
FAIR TRADE, SUSTAINABILITY, AND SOCIAL CHANGE

Andrés Rivarola Puntigliano and José Briceño-Ruiz (*editors*)
RESILIENCE OF REGIONALISM IN LATIN AMERICA AND THE CARIBBEAN
Development and Autonomy

Godfrey Baldacchino (*editor*)
THE POLITICAL ECONOMY OF DIVIDED ISLANDS
Unified Geographies, Multiple Polities

Mark Findlay
CONTEMPORARY CHALLENGES IN REGULATING GLOBAL CRISES

Nir Kshetri
CYBERCRIME AND CYBERSECURITY IN THE GLOBAL SOUTH

Kristian Stokke and Olle Törnquist (*editors*)
DEMOCRATIZATION IN THE GLOBAL SOUTH
The Importance of Transformative Politics

Jeffrey Wilson
GOVERNING GLOBAL PRODUCTION
Resource Networks in the Asia-Pacific Steel Industry

International Political Economy
Series Standing Order ISBN 978–0–333–71708–0 hardcover
Series Standing Order ISBN 978–0–333–71110–1 paperback
(*outside North America only*)

You can receive future titles in this series as they are published by placing a standing order. Please contact your bookseller or, in case of difficulty, write to us at the address below with your name and address, the title of the series and one of the ISBNs quoted above.

Customer Services Department, Macmillan Distribution Ltd, Houndmills, Basingstoke, Hampshire RG21 6XS, England

Climate Change and Global Policy Regimes

Towards Institutional Legitimacy

Edited by

Timothy Cadman

Research Fellow, Griffith University, Australia

palgrave
macmillan

First published 2013 by
PALGRAVE MACMILLAN

Palgrave Macmillan in the UK is an imprint of Macmillan Publishers Limited, registered in England, company number 785998, of Houndmills, Basingstoke, Hampshire RG21 6XS.

Palgrave Macmillan in the US is a division of St Martin's Press LLC, 175 Fifth Avenue, New York, NY 10010.

Palgrave Macmillan is the global academic imprint of the above companies and has companies and representatives throughout the world.

Palgrave® and Macmillan® are registered trademarks in the United States, the United Kingdom, Europe and other countries.

ISBN 978–1–137–00611–0

A catalogue record for this book is available from the British Library.

A catalog record for this book is available from the Library of Congress.

10 9 8 7 6 5 4 3 2 1
22 21 20 19 18 17 16 15 14 13

MIX
Paper from responsible sources
FSC® C013604

For Audrey

Contents

Figures and Tables

Figures

Tables

Preface

In 2001, I was engaged as a research consultant by the international environmental organisations Greenpeace and the World Wide Fund for Nature (WWF) to undertake field research into the establishment of tree plantations to be used as carbon 'sinks' to absorb greenhouse gas emissions. This idea had arisen from the intergovernmental climate change talks and was hailed a 'win-win' solution to both encourage sustainable forestry and reduce carbon dioxide in the atmosphere. One project, a joint venture in Tasmania between a forestry corporation and a Japanese utility company, was at that time being promoted as a 'prototype' carbon fund, aimed at encouraging financial markets to invest in carbon 'offsets'. Unfortunately, however, ancient native eucalypt forests and rainforests had been cleared to make way for monoculture tree farms of exotic species, and the once pristine forests they replaced – with all their potential to store carbon – had been cleared and burnt.

Having prepared my report, I made my way to the climate negotiations in The Hague, and what a bizarre experience they were. Outside the convention centre was a giant wooden ship, built by environmental campaigners, symbolising that the venue would soon be under water if governments failed to take action to combat climate change. Hundreds of protesters milled around, waving placards and shouting slogans, held back by Dutch police. Inside the talks, carbon sinks had become the *cause célèbre* of discussions, with the United States lobbying hard behind closed doors to have forests included as means of offsetting their emissions, and civil society delegates strongly arguing the opposite in the corridors, to whoever would listen. Unexpectedly, my research findings briefly became the centre of media attention. The talks themselves collapsed without agreement, partly around the very issue of sinks, and had to be rescheduled. Back in Australia, the Senate conducted an inquiry into the effectiveness of Australia's response to global warming, including the role of forestry investment schemes. As a result of the recommendations in the inquiry, the Australian Securities and Investments Commission made a number of rulings about the disclosure statements and claims that could and could not be made by companies promoting the potential of tree plantations to sequester carbon. The Tasmanian case study revealed that an idea that may look good in theory can have disastrous consequences on the ground if appropriate measures are not put in place first, and all the necessary expertise and advice brought to bear. In a sense, it represents the challenges confronting the governance of climate change management in microcosm.

This book is a collaborative effort between research colleagues from Australia and around the world. In the chapters contained in this volume, we have attempted as comprehensive an investigation as possible of the challenges confronting the governance of climate change management. We have not been able to cover every issue. Some people may be disappointed not to find a detailed analysis of climate science itself, or the environmental consequences of climate change. We believe that these issues have been more than adequately covered by many scholars before and that the reality of climate change is undeniable. We have chosen instead to look at the human dimensions of climate change, and how these are played out within a range of institutional contexts. In order to provide some consistency in what is a very broad scope, we have used the 'Cadmanian framework' contained in my first book, *Quality and legitimacy of global governance: case lessons from forestry.* It is included in the Introduction to this volume. Each contributor has responded to the approach presented there in their own way, as it has spoken to the situations they have written about. I would refer readers to that book and the Doctoral Thesis upon which it was based if they are looking for more detail. Any theoretical failings are my own, and not those of my colleagues, who have so kindly agreed to be tarred with the same brush.

Timothy Cadman
Nathan,
Queensland,
A.M.D.G

Acknowledgements

Our thanks go to all those who have helped with this book in any way. Professor Timothy Shaw, Christina Brian and everyone at Palgrave Macmillan have been as helpful and encouraging as ever. We would particularly like to acknowledge the University of Southern Queensland, Faculty of Business and Law Research Scholars Scheme. This enabled the convening of the 'Climate Governance Symposium: Issues, Interests and Institutions in Global Policy Regimes', held in Toowoomba 2–4 March 2011, which provided the opportunity for a number of the contributors to this volume to meet with each other, share their research and receive constructive feedback. We greatly appreciate the mentoring and support provided by Dr Peter Phillips, the academic respondents and the anonymous reviewers; without their help the event would not have been as intellectually rigorous and rewarding as it was. For the wood fibre used in the production of this work, we salute the trees.

Contributors

Andrea C. Berringer is currently an independent researcher. She is a graduate of the UNU Environment and Human Security Programme (2010) and Oxford's Refugee Studies Centre Summer School (2009). Andrea is an associate with the Millennium Alliance for Humanity & the Biosphere (MAHB) and a member of the Roster of Experts, Climate and Development Knowledge Network (CKDN). She specialises in governance responses to human migration, specifically responses to displacement due to climate change processes.

Timothy Cadman is a research fellow in the Key Centre for Ethics, Law, Justice and Governance, and the UNU Institute for Ethics, Governance and Law at Griffith University, Queensland, Australia. He is also a research fellow in the international Earth Systems Governance Project and specialises in the governance of sustainable development, natural resource management, climate change and forestry, and responsible investment. His first book, *Quality and legitimacy of global governance: case lessons from forestry,* was published by Palgrave Macmillan in 2011.

Geoff Cockfield is Associate Professor in Politics and Economics and Research Associate in the Australian Centre for Sustainable Catchments at the University of Southern Queensland, Australia. His research focuses on natural resources management in rural landscapes and climate change policy.

Julie Cotter is a professor in the Faculty of Business and Law and the Acting Director of the Australian Centre for Sustainable Business and Development at the University of Southern Queensland, Australia. Professor Cotter is highly regarded for her research in corporate reporting and governance. She conducts leading-edge international research into carbon disclosure and leads research being undertaken by a team of researchers and industry professionals with expertise in carbon management.

Lauren E. Eastwood is Associate Professor of Sociology at the State University of New York, College at Plattsburgh, USA. Her research interests include ethnographic analyses of UN environmental policymaking and community responses to energy extraction in North America. She was a 2009–2011 Abe Fellow with the Social Science Research Council. She is the author of *The social organization of policy: an institutional ethnography of UN forest deliberations* (2005) and *Negotiating the environment: civil society, globalisation, and the UN* (2013).

Fred Gale is Senior Lecturer in Politics and International Relations in the School of Social Sciences, University of Tasmania, Australia. He is a member of the Faculty of Arts Governance and Implementation Research Group and a member of the Australian National University, Asia-Pacific Network for Environmental Governance. He conducts research on governance theory and practice, commodity certification, forest politics, ecological political economy and sustainable consumption and production.

Guilherme Gonçalves is Executive Manager and Founding Partner at Limpgas Tecnologia, a Brazilian startup pioneer in the development of innovations for the reduction and treatment of greenhouse gases and toxic effluents and residue by physical processes. He also develops new methodologies for the market entry of green innovations.

Jeff Gow is Professor of Economics in the School of Accounting, Economics and Finance in the Faculty of Business and Law at the University of Southern Queensland, Toowoomba, Australia. He is also a research associate of the Health Economics and HIV/AIDS Research Division (HEARD), University of KwaZulu-Natal, Durban, South Africa. His research interests encompass health economics, especially HIV/AIDS globally, and Australian agricultural economics and policy.

Matthew Haigh is a semiotician who writes about financial markets and ecology. He teaches at the School of Oriental and African Studies at the University of London, UK, and has also taught at the universities of Aarhus, Toulouse and Amsterdam. He is a qualified chartered accountant and certified systems auditor and has a professional background in auditing. He serves as founding editor of the *Journal of Sustainable Finance & Investment.*

Richard Hil is Honorary Associate at the University of Sydney, Centre for Peace and Conflict Studies and Adjunct Associate Professor, School of Social Work and Human Services, Griffith University, Australia. His research and teaching interests are in the areas of social and community development, criminology, youth studies and peace and conflict studies. Richard is author of several books, including the recent publications *Erasing Iraq* (2010), *Surviving care* (2010) and *Whackademia: an insider's account of the troubled university* (2012).

Guilherme B.R. Lambais is a researcher on the economics of technology and agriculture from an environmental perspective. He has been a researcher at the University of Texas at Austin and the Center for Agricultural and Environmental Research at the University of Campinas, where he was editor of *Leituras de Economia Política*, a graduate student journal. He also participates in business venture creation, translating theory into practice, and to date has provided advice on two green startups.

Tek Narayan Maraseni is Deputy Director (Operations) in the Australian Centre for Sustainable Catchments at the University of Southern Queensland, Australia. He is also Visiting Professor with the Chinese Academy of Sciences (Lanzhou). He is working in the areas of carbon accounting and modelling. Over the last seven years, he published over 75 works, including two books. His publications have been acknowledged through the receipt of a number of national and international fellowships, grants and awards.

Jamie Pittock is a research fellow in the Crawford School of Public Policy at the Australian National University, Canberra, Australia. He is Programme Leader for the Australia and United States' Climate, Energy and Water Nexus project of the US Studies Centre and ANU, as well as Director of International Programmes for the United Nations Educational, Scientific and Cultural Organisation (UNESCO) Chair in Water Economics and Transboundary Water Governance. His research focuses on how society can better sustain water and energy supplies while conserving biodiversity and responding to climate change.

Nick Rose is a research assistant at Griffith University's School of Environment, investigating the role that urban agriculture plays in climate change adaptation, building urban resilience and addressing future food security, focusing on practice, research and policy innovations in Melbourne. A Director of the Food Connect Foundation in Brisbane and national coordinator of the Australian Food Sovereignty Alliance, his recently completed doctoral research examined the emergence of food sovereignty as a transformative social movement.

Chris Taylor is a research fellow in the Melbourne Sustainable Society Institute at the University of Melbourne, Australia. He is Research Director of the Land Use, Forestry and Agriculture Plan, a joint project between Beyond Zero Emissions and the University that aims to develop a road map for the transition to a de-carbonised Australian economy on a ten-year time scale. The Plan will be released in March 2013.

Heather Zeppel is Associate Professor and Mid Career Research Fellow in the Australian Centre for Sustainable Business and Development at the University of Southern Queensland, Australia. She was previously senior lecturer in sustainable tourism at James Cook University in North Queensland, Australia. Her research interests include carbon mitigation by local government, climate change and tourism, environmental sustainability, ecotourism and Indigenous tourism. She is the author of *Indigenous ecotourism: sustainable development and management* (2006).

Abbreviations

ACCA	Association of Chartered Certified Accountants
ADB	Asian Development Bank
AIDS	See HIV/AIDS
AWG-KP	Ad Hoc Working Group on Further Commitments for Annex 1 Countries under the Kyoto Protocol (UNFCCC)
AWG-LCA	Ad Hoc Working Group on Long-Term Cooperative Action under the Convention (UNFCCC)
APP	Asia Pacific Partnership
AOSIS	Alliance of Small Island States
ARB	Autonomous Region of Bougainville
ART	Antiretroviral Treatment
BASIC	Brazil, South Africa, India and China
BINGO	Business International NGO
BRIC	Brazil, Russia, India, China
CAN	Climate Action Network
CBD	United Nations Convention on Biological Diversity
CCE	Chicago Climate Exchange
CCP	Cities for Climate Protection (ICLEI)
CCRF	Climate Change Reporting Framework (CDSB)
CCS	Carbon Capture and Storage
CDM	Clean Development Mechanism (UNFCCC)
CDP	Carbon Disclosure Project
CDSB	Climate Disclosure Standards Board
CER	Certified Emissions Reduction (CDM)
CFS	Committee on World Food Security
COP	Conference of Parties (UN)
CSD	United Nations Commission on Sustainable Development
CSM	Civil Society Mechanism (CFS)
CSO	Civil Society Organisation
CSW	Commission on the Status of Women (UN)
DALY	Disability-Adjusted Life Year
DCCEE	Department of Climate Change and Energy Efficiency
DNA	Designated National Authority (CDM)
DOE	Designated Operational Entity (CDM)

EB	Executive Board (CDM)
ECOSOC	Economic and Social Council of the United Nations
ECT	Energy Charter Treaty
EJF	Environmental Justice Foundation
ENGO	Environmental NGO
ETS	Emissions Trading Scheme
EU	European Union
EUAMD	European Accounts Modernisation Directive
FAO	Food and Agriculture Organisation
FCCC	See UNFCCC
FSC	Forest Stewardship Council
G7	Group of Seven
G8	Group of Eight
GCM	General Circulation Model
GEF	Global Environment Facility (UN)
GHG	Greenhouse Gas
GHI	Global Health Initiative
GM	Genetically Modified
GPC	Global Protocol for Community (ICLEI)
GRI	Global Reporting Initiative
GSF	Global Strategic Framework (FAO – WFS)
HadCM3	Hadley Centre Coupled Model, Version 3
HadOCC	Hadley Centre Ocean Carbon Cycle Model
HCFC	Hydrochloroflourocarbon
HFC	Hydroflourocarbon
HIV/AIDS	Human Immunodeficiency Virus Infection/Acquired Immunodeficiency Syndrome
HLPE	High Level Panel of Experts (CFS)
IAASTD	International Assessment of Agricultural Knowledge, Science and Technology for Development
IAEA	International Atomic Energy Agency
ICLEI	International Council for Local Environmental Initiatives
IDP	Internally Displaced Person
IEA	International Energy Agency
IEAP	International Local Government Greenhouse Gas Emissions Analysis Protocol (ICLEI)
IEF	International Energy Forum
IET	International Emissions Trading (UNFCCC)
IFAD	International Fund for Agricultural Development
IGO	Intergovernmental Organisation
IIRC	International Integrated Reporting Committee

IMF	International Monetary Fund
INC/FCCC	Intergovernmental Negotiating Committee for a Framework Convention on Climate Change (UNFCCC)
INCR	Investor Network on Climate Risk
INGO	International NGO
IOM	International Organisation for Migration
IPCC	Intergovernmental Panel on Climate Change (UNFCCC)
IPEEC	International Partnership on Energy Efficiency Cooperation
IPO	Indigenous Peoples' Organisation
IRENA	International Renewable Energy Agency
ITL	International Transaction Log (CDM)
JI	Joint Implementation (UNFCCC)
JLG	Joint Liaison Group (UN)
JWP	Joint Work Plan (CBD and Ramsar)
KP	Kyoto Protocol (UNFCCC)
KPAF	Kyoto Protocol Adaptation Fund (UNFCCC)
LDC	Least Developed Countries
LDCF	Least Developed Countries Fund (UNFCCC)
LG	Local Government
LGMA	Local Government and Municipal Authorities
LULUCF	Land Use, Land Use Change and Forestry (UNFCCC)
MDG	Millennium Development Goals (UN)
MEA	Multilateral Environment Agreement
MEF	Major Emitters Forum
MOP	Members of Parties (UN)
MRV	Monitoring Reporting and Verification (UNFCCC)
MSM	Men who have sex with men
NAMA	Nationally Appropriate Mitigation Action (UNFCCC)
NAP	National Adaptation Plan (UNFCCC)
NAPA	National Adaptation Programmes for Action (UNFCCC)
NGO	Non-governmental Organisation
OECD	Organisation for Economic Cooperation and Development
OAPEC	Organisation of Arab Petroleum Exporting Countries
OLADE	Latin American Energy Association (Organización Latinoamericana de Energía)
OPEC	Organisation of Petroleum Exporting Countries
PDD	Project Design Document (CDM)
PEPFAR	President's Emergency Plan for AIDS Relief

PIN	Project Identification Note (CDM)
PNG	Papua New Guinea
PPP	Public Private Partnership
PRI	Principles for responsible Investment (UN)
PSIDS	Pacific Small Island Developing States
REDD	Reducing Emissions from Deforestation and Forest Degradation (UNFCCC)
REDD+	See REDD
RINGO	Research and Independent NGO
SBI	Subsidiary Body for Implementation (UNFCCC)
SBSTA	Subsidiary Body for Scientific and Technical Advice (UNFCCC)
SBSTTA	Subsidiary Body on Scientific Technical and Technological Advice (CBD)
SEC	Securities and Exchange Commission
SSPS	Statistical Package for the Social Sciences
STRP	Scientific and Technical Review Panel (Ramsar Convention)
TRIFFID	Top-down Representation of Interactive Foliage and Flora Including Dynamics
UN	United Nations
UNCED	United Nations Conference on Environment and Development
UNCHE	United Nations Commission on the Human Environment
UNCTAD	United Nations Conference on Trade and Development
UNCTC	United Nations Centre for Transnational Corporations
UNDP	United Nations Development Programme
UNEP	United Nations Environment Programme
UNEPFI	United Nations Environment Programme Finance Initiative
UNFCCC	United Nations Framework Convention on Climate Change
UNHCR	United Nations High Commissioner for Refugees
UNOCHA	United Nations Office for the Coordination of Humanitarian Affairs
VCO	Voluntary Carbon Offset
WBCSD	World Business Council for Sustainable Development (formerly Business Council for Sustainable Development)
WCED	World Commission on Environment and Development

WEDO	Women's Environment and Development Organisation
WEF	World Economic Forum
WFS	World Food Summit (FAO)
WHO	World Health Organisation
WMCCC	World Mayors Council on Climate Change
WRI	World Resources Institute
WTO	World Trade Organisation
WWF	World Wide Fund for Nature (World Wildlife Fund)

Introduction: Global Governance and Climate Change

Timothy Cadman

Purpose of this book

In this volume of edited chapters, the contributors evaluate how the various institutional arrangements, actors and agendas that comprise what has been referred to as the global climate regime complex impact governance quality. Considering the fundamental features of the climate regime complex – notably interests and power – governing the management of climate change is a very fragmented affair. It has been further argued that this fragmentation might actually have advantages over other formations, particularly with regard to adaptability and flexibility – but only if the right conditions are in place (Keohane and Victor 2010: 25). Using an approach based on governance analysis, this book explores these conditions to determine the institutional legitimacy of contemporary responses to anthropogenic climate change.

Institutions and instruments in the governance of climate change management

The role institutions play in tackling the problem of climate change and how they help or hinder the actions taken have become increasingly significant in recent years (Barnett 2010: 316). The establishment of the Intergovernmental Panel on Climate Change (IPCC) in 1991 set the intellectual framework for global climate deliberations. Once the United Nations Framework Convention on Climate Change (UNFCCC) came into force in 1994, IPCC's largely technocratic approach was heavily influenced by the neo-liberal market order, exemplified by three flexibility mechanisms of the Kyoto Protocol (KP): International Emissions Trading (IET), Joint Implementation (JI) and the Clean Development

Mechanism (CDM). Since 2005 there has also been a European Union Emissions Trading Scheme (EU ETS), linked since 2008 to the purchase of carbon credits from JI and CDM projects. Between the Conference of Parties (COP) 7 in Marrakesh in 2001 and COP 12 in Nairobi in 2006, the original IPCC focus on mitigating atmospheric processes and mean global temperatures began to be supplanted by discussions around vulnerability and adaptation issues (Bäckstrand and Lövbrand 2007: 126–130). Although advocates for environmental justice and southern countries had argued on this perspective for some time previously, the eventual policy shift led to climate change no longer being simply framed as an environmental problem, but as a societal problem with an environmental component (Barnett 2010: 15). This has resulted in a growing interest in strengthening the capacity of southern, developed countries to respond to climate change. Although this has not greatly affected developed countries' actions regarding implementation of the KP, it has added further uncertainty regarding the shape of the post-2012 climate policy landscape (Bäckstrand and Lövbrand 2007: 128–129).

The problem of climate change governance is made more difficult by the ever-increasing and/or evolving numbers of new bodies, mechanisms and proposals relating to policy development and implementation. Climate change management constitutes one of the most significant and normatively embedded post-1992 Rio 'Earth' Summit 'meta' institutions, and the UNFCCC has its own sets of institutional arrangements, replete with a variety of governance systems to tackle the problem of climate change through market-based or other sustainable development initiatives. However, multilateral environmental agreements are no longer the only, or even central, mechanism for global environmental governance and they sit alongside a range of other forms, both public and private, functioning on many different levels and layers of authority and including many different actors – not just governments (Andonova et al. 2009: 52). Although it is mostly state actors who exercise authority on the basis of their control at the national level, climate change governance is simultaneously global and local, state and non-state, and it is characterised by the existence of many forms of authority through which different constellations of actors interact to shape policy outcomes (ibid: 67; Betsill et al. 2006: 141). Consequently, understanding governance more in terms of multi-level, multi-spatial networks is likely to better assist in determining how negotiations between national governments and non-state stakeholders can more effectively contribute to sustainable development post-2012 (Bumpus and Cole 2010: 543). Governance analysis has therefore seen a move away from international relations as the sole lens through which to examine the phenomenon, and comparative politics, political economy and a range of political science approaches are being applied to interpret and shape developments (Bulkeley 2010: 311).

Interests and issues in the governance of climate change management

The substantive output of Rio, *Agenda 21*, embedded non-state participation in the normative framework of international environmental policy and the role of non-state interests in environmental decision-making at all levels was formally acknowledged (United Nations 1992: 230–235). The expectation for increased citizen participation in decision-making that these developments have brought about therefore raises some dynamic tensions between state and non-state actors in the creation of global environmental policy. A range of participants are involved in the governance of climate change management, from both the geo-political and sectoral spheres, from the local through to the international, each of which is influencing the other (Bulkeley 2010: 312). Local government, for example, is taking initiatives in its own right and within national and international arenas (Andonova et al. 2009: 52–53; Koehn 2010: 405). The trend is now for transnational cooperation across governmental levels, regions and within networks that include non-governmental organisations (NGOs) and corporations. Non-state actors, previously outside the formal decision-making arenas, are now playing a role in the formation of public policy, and their participation is challenging traditional conceptions of power and authority (Andonova et al. 2009: 52–53). This has implications for the nature of relations between state, society and the economy, and for previous notions of legitimacy and accountability (Bulkeley 2010: 312). Given the cross-border nature of the issue, its complexity and the number of players involved, climate change policymaking has lent itself to this type of governance. The KP in particular has opened up climate change to market mechanisms, creating governance structures that require cooperation between state and non-state actors, even if it is the nation state that ultimately endorses them (Andonova et al. 2009: 57–58).

Sustainable development has been identified as the bridging mechanism for integrating climate change and development policies. This nexus will be highly important in the post-2012 agreement, especially in the provision of financial and technical support (Metz and Cox 2008: 99–100). This is especially the case for the emerging 'Post-Kyoto' mechanism to reduce emissions from deforestation and forest degradation (REDD or REDD+). Following the Rio model of engagement, such mechanisms allow for multi-stakeholder participation in the governance of climate change management and enhance local capabilities. However, current arrangements are providing the West with a cheap way to avoid taking action itself, while at the same time privatising the global commons – and making nature the subject of private property rights through the market. These realities reinforce the domination of rich countries over poor, while alienating the poor from their land and decision-making over common resources (Okereke 2010: 470).

There is consequently an inherent and ongoing tension between sustainable development and the governance of climate change management. Contradictions are evident in the KP and the CDM in particular. On the sustainable development side of the equation it promotes specialisation and North/South technology transfer and provides an efficient way of tackling emissions reduction cost-effectively. Developing countries argue that developed countries should take the lead in reducing emissions; developed countries in return argue that they were unaware of the impacts of emissions previously and that it is fairer to allocate burden sharing on the basis of current emission levels (Okereke 2010: 470). The contradictions surrounding sustainable development and North/South relations reveal underlying geo-political tensions about global governance and climate change management. It is usually the case that the more powerful the country, the greater influence it has on international policy positions that best suits it (Okereke 2010: 464). The current regulatory carbon-related framework and its North to South market-based models of capital transfer and carbon finance mechanisms have been characterised as distributing development unevenly. It has even been claimed that the management of sinks through the CDM represents the 're-territorialization' of Northern control over the South. Whoever has the responsibility and authority over 'sinks spaces' – specifically plantations and the management of trees – is especially relevant and is creating a form of global environmental governance that is effectively imperial, an 'empire of carbon management and control' (Paterson and Stripple 2007: 163).

Nevertheless, the formation of intergovernmental climate change policy has also been influenced by a reformist movement led by NGOs and small island states, particularly in the case of Kyoto after 2001, and once the United States withdrew from negotiations. Greater attention is now being given to participation, accountability and transparency in the implementation of Kyoto mechanisms, which has led to some increase in support for carbon markets, voluntary partnerships and ETS among previously hostile stakeholders. This in turn has resulted in greater alignment with the agenda for North/South equity in climate policy. However, whether new governance practices that balance sustainable development, market efficiency and North/South equality will emerge remains to be seen (Bäckstrand and Lövbrand 2007: 136–141).

Responses beyond the climate regime

As the discussions above have indicated, there is no single global climate change regime. Rather, the principles, norms, rules and decision-making procedures are spread across a wide range of initiatives operating at the international, plurilateral, bilateral and national levels and promoted by both public and private actors. While there is a tendency to focus on intergovernmental policy responses, these constitute only some of the elements of the broader regime complex. Important plurilateral initiatives include the Asia Pacific Partnership (APP) and the Major Emitters Forum (MEF). Bilateral efforts include collaboration by the United Kingdom and China on technologies for coal combustion and the United States and India on nuclear power (Keohane and Victor 2010: 3–8). In addition, there are a range of sub-national, market and civil society initiatives, including the Chicago Climate Exchange (CCE), the International Council for Local Environmental Initiatives (ICLEI) campaign Cities for Climate Protection (CCP) and standards for voluntary offset projects, such as the World Wildlife Fund's Gold Standard.

Other critical policy areas are impacted by climate change, one of the most significant being human services. In the case of health, the well-being of millions of people will be affected by extreme weather events, changes in the spatial distribution of infectious diseases and the ability to maintain key services such as public health and health care, education and health-related infrastructure (Sterling 2010: 20–21). The World Health Organisation (WHO), the coordinating public authority for global health, has already estimated, using available data commencing in the 1970s, that global climate change-related deaths had reached 150,000 per annum by 2000 (WHO in Patz et al. 2005: 13). During roughly the same period, a nexus between climate change and migration was observed by a number of commentators, with the term environmental refugee first appearing in the 1970s, a term reinforced by further studies in 1985 and again in 1995. The studies estimated the number of refugees to be at 25 million, but it could rise to as many as 200 million by 2025 (Marquina 2010a; 2010b: 192–193). Despite these predictions, policy response to date has been labelled insufficient and the governance of migration incoherent (ibid: 203; Betts and Loescher 2010: 1). Water availability, both in terms of quality and quantity, is a further predicted consequence of climate change, and impacts on agriculture and food availability for local populations are seen as being particularly pressing. Addressing water-related legal, regulatory and institutional arrangements is therefore essential, and some policy reform is already underway (Mustafa 2010: 35–43). In Europe, preparing for water scarcity is an acknowledged element of adaptation to climate change and a component in the European Union's latest White Paper, which is itself a follow-up to a 2007 Green Paper, where it was acknowledged that changes in temperature of between 2°C and 3°C could affect the water resources of as many as three billion people (Monreal and Amelin 2010: 70–74).

These and other issues have made climate change a major human security priority for the international community. But although everybody appears to accept this is a looming problem, there is at present no consensus on how to deal with it. This is not only partly due to issues of complexity, as every country will have different local manifestations, but also due to divergent agendas on climate change-related factors such as greenhouse gas (GHG) emissions, deforestation and environmental degradation generally, and economic growth (Caballero-Anthony 2010: 393). Nevertheless, there is recognition that one response arising out of existing mitigation and adaptation policy approaches is that of conflict prevention. Here the discussion regarding climate change inevitably cycles back to governance, particularly in terms of responsibility, decision-making and problem solving, all of which are identified as critical determinants of institutional legitimacy (German Advisory Council on Global Change 2008: 167 and 183).

Evaluating institutional governance legitimacy: developments in research and analysis

While the necessity of mitigating and adapting to the impacts of climate change has been debated at length, this has generally not been the case for the institutions themselves that are charged with governing the formation of the necessary policy responses in an effective and responsible way (Thynne 2008: 327–329). Furthermore, despite the many dimensions of climate change being tackled at the global policy level, there continues to be no integrating approach to governance.

The problem is made more difficult by the institutional framework at the global level, which is often compartmentalised, whereas climate change is a cross-sectoral issue. Concerns about institutional effectiveness have been voiced at numerous negotiations, and there have been widespread calls for reform of the existing governance arrangements for both climate change management and sustainable development generally. But the development of international environmental policy still does not take place in an arena where all ideas are welcomed and adopted. Nevertheless, there has been a shift in the international climate regime towards a greater recognition of norms and principles that stress standards of behaviour, although justice and equity remain contested terms within the international climate regime (Okereke 2010: 463). One of the key governance challenges is therefore how to design policies that combine normative expectations of 'good' behaviour with conventional power politics (ibid: 471).

There is a growing recognition that more research is needed to explore the dimensions of governance quality in climate mechanisms (Corbera and Schroeder 2011; Thompson et al., 2011). A second, equally important, and related observation is that greater attention should be paid to evaluating the success of climate change policies on the basis of the social processes that drive decision-making (Barnett 2010). This is all the more pressing in view of the fact that governance, as opposed to government, is increasingly acknowledged as a primary means by which social and political interaction can be understood in the global context of state, society and the market. This perspective is predicated on recognising the significance of the social–political nature of stakeholder interactions within contemporary governance systems and the structures and processes that underpin them (Kooiman 2000). This grounds theory and practice within the normative assumption that structures and processes are fundamental to understanding the quality of contemporary governance.

Concerns about the quality of climate change governance often centre upon gaps in legitimacy, and not just specific institutional or technical aspects, particularly among developing country stakeholders (Streck et al. 2009). It has been argued that 'a greater degree of self-awareness might help us to ascertain just what it is we are doing when we use a word like legitimacy, and help us see more clearly what changes this betokens in the discourses of international politics' (Mulligan 2006: 375). Legitimacy has become a core analytical problem for governance scholars, but its study is still in its infancy (Biermann and Gupta 2011; Biermann et al. 2009). Two theories currently dominate.

Legitimacy can be 'input oriented': that is, derived from the consent of those being asked to agree to the rules and concerning such procedural issues as the democratic arrangements underpinning a given system. Legitimacy can also be 'output oriented': derived from the efficiency of rules, or criteria for 'good' governance, and demonstrated by substantive outcomes (Kjaer 2004). Output-oriented legitimacy can be achieved in climate mechanisms, but a higher degree of input-oriented legitimacy is also necessary, which may require a trade-off between the two (Lederer 2011). It is only through significant interaction that stakeholder interests can be aligned from the local community level to international negotiations. The best way to enable this alignment is to examine how stakeholder interaction is facilitated in climate processes (Thompson et al. 2011). Legitimacy is framed quite specifically in the approach adopted in this volume, as it is conceived as the end point of activity within an institution. It is determined by the degree of successful interaction between the structural and procedural components of the institution's governance system, that is, an integrating model, as depicted in Figure I.1.

The institutional arrangements, underpinning the interactions between the various participants in policy regimes, also have a bearing on governance quality (Koenig-Archibugi 2006: 24). These arrangements refer to commonly identified attributes such as interest representation, decision-making and implementation. One of the major problems with

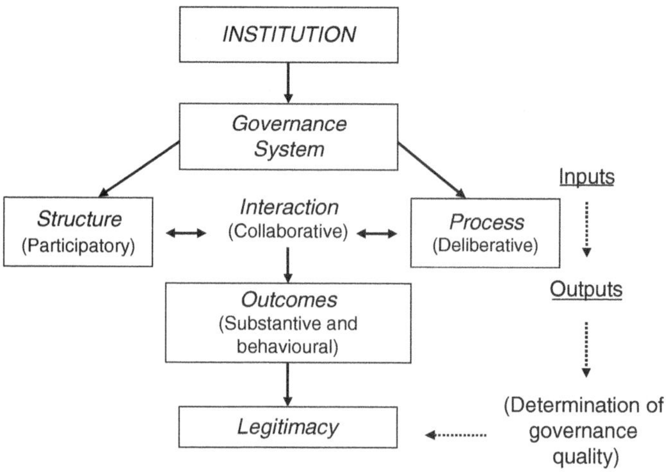

Figure I.1 Theoretical model for evaluating contemporary global governance
Source: Cadman (2011: 5)

studies that examine governance quality is that the attributes chosen for investigation are often limited in number or quite arbitrary and not always reflective of the whole suite of arrangements necessary for investigating institutional performance as a whole. The current focus (almost obsession) on accountability and transparency is a good example. Recent work on the quality and legitimacy of global governance addresses these criticisms through the development of a comprehensive analytical approach based on two core governance arrangements: structure and process. Here, contemporary governance is conceived in terms of 'participation as structure' and 'deliberation as process' (Cadman 2011: 4–5). This is an extension of the existing idea that governance is now to be understood in terms of both structure and process (Pierre and Peters 2000: 14). In such a context, participation and deliberation have a functional significance beyond their particular expression in a given institution; it is not the institution per se but rather how participation and deliberation occur within it that determines the effectiveness of its governance. In totality, the interactions within a given institution represent the major components of what can be termed governance systems, and what structures and processes these systems utilise yield important information about the capacity of a specific institution under investigation to combat climate change.

The structural features of governance focus on which actors are viewed as valid participants. In contrast, process requirements focus on the means employed to reach decisions and implement them. Two principles have been created to emphasise those normative values underpinning participation and deliberation: participation is expected to be *meaningful* (that is, that involvement is genuine rather than tokenistic); deliberation is expected to be *productive* (that is, that discussion and dialogue are fruitful and actually deliver outputs that can be acted upon). Based on this division between structure as participation and process as deliberation, the meaning of these two principles is elaborated by developing criteria and indicators to examine the degree to which they are achieved in a given institutional policy context. Principles and criteria are not usually capable of being measured directly, but they are formulated to provide a determination on the degree of compliance. They are consequently linked to *indicators* that are hierarchically lower, represent quantitative or qualitative parameters, and describe conditions indicative of the state of the governance system as they relate to the relevant criterion. The intention behind the placement of these attributes within such a framework is to ensure that they are located at the right level, to allow for a consistent top–down analysis from principles, to criteria, and subsequently to indicators. Consistency in this context relates to the correct location within the framework: it is important that elements are placed at the appropriate level and do not overlap or duplicate those at another and are linked back to the appropriate parameter at a higher level (Lammerts van Bueren and Blom 1997).

Meaningful participation is demonstrated through two criteria: *interest representation* and *organisational responsibility*. Interest representation has been linked to three elements of governance that function on the indicator level: *inclusiveness*, demonstrating who participates in a governance system; *equality*, indicating the nature of the relationship between participants; and *resources*, referring to the economic, technical or institutional capacity of participants to represent their interests within the system. Organisational responsibility comprises two indicators: *accountability* and *transparency*. These indicators, usually treated together in the literature, refer to the extent to which the behaviour of participating organisations can be called to account both inside the institution and externally by the public at large, as well as being visible or open to scrutiny by other actors within the institution and beyond. Productive deliberation is demonstrated through two criteria: *decision-making* and *implementation*. Three indicators are linked to decision-making: *democracy*, referring not to a specific mode of democracy but rather to the extent to which a system can be deemed to be functioning democratically; agreement, referring to the method in which decisions are reached (for example voting or consensus); and *dispute settlement*, indicating the system's capacity to manage conflict when there is no agreement or there are challenges to decisions made. Another three indicators are linked to implementation: *behaviour change*, used to determine whether the implementation of agreements or substantive outcomes results in changed behaviour regarding the problem that the system was created to address; *problem solving*, referring to the extent to which the system has solved the problem it was created to address; and *durability*, capturing the two related elements of adaptability and flexibility, as well as longevity (Table I.1).

It should be noted that the key governance concept of legitimacy, identified by many scholars, is not directly included, as it is understood, as Figure I.1 demonstrates, as the end point of activity within the institution. The normative concept being stressed is that the ends and means are equally important. Both are related and consequential to one another and both play a role in legitimacy.

In this book, 'good' governance is therefore not attributed to any single institutional arrangement, such as accountability or transparency

Table I.1 Hierarchical framework for the assessment of governance quality

Principle	**Criterion**	Indicator
'Meaningful participation'	*Interest representation*	Inclusiveness Equality Resources
	Organisational responsibility	Accountability Transparency
'Productive deliberation'	*Decision-making*	Democracy Agreement Dispute settlement
	Implementation	Behavioural change Problem solving Durability

Source: Cadman (2011: 17)

and so forth, even though these are of course important. Rather than evaluating the performance of an institution on the basis of a few individual attributes, the approach adopted consequently looks at institutional governance at a systemic level. This provides important information concerning some of the broader parameters affecting quality of governance and their impact on policy responses to climate change.

Content and key findings

It could be argued that the approach adopted in this study is partly constructivist in nature, particularly regarding the behaviour of actors within institutional venues (Haas 2002: 74). This perspective shares an interest with state-centric regime theory regarding the effectiveness of global environmental governance, particularly as these relate to intergovernmental organisations (Koenig-Archibugi 2006: 3–12). However, the case studies in this volume examine a wider range of institutional types and systems, both state and non-state, and are not confined to any one critical approach, state-centric or otherwise.

The first four chapters focus on some of the 'big picture' debates associated with climate change. *Chris Taylor* explores how the various discourses within the governance of climate change management arose, and what this has meant for various stakeholders. He finds that once responsibility for climate change management moved from the scientific to the government realm, it became elitist and exclusionary. *Fred Gale* explores the limits of the intergovernmental approach to the governance of climate change management. He identifies inadequate structures and processes for meaningful participation and productive deliberation and argues that these deficiencies have led to the current negotiating impasse. Non-state actors, in particular, are unable to significantly influence outcomes. Effective change is more likely to be achieved by refocusing pressure directly on market actors than on attempts to control them via intergovernmental regulation. Focusing specifically on gender as a 'stakeholder' in climate deliberations, *Lauren Eastwood* elaborates how the ideological and rhetorical components associated with both gender and climate change are used strategically by non-state actors as they attempt to represent their interests in new ways and influence policymaking processes. Adopting a state-centric perspective, *Geoff Cockfield* investigates the controversy around the merits of adaptation to, or mitigation of, climate change. He predicts that liberal-democratic states will struggle to accommodate the demands of those who want climate change policy to be focused on mitigation. As a result, state responses will include non-decision-making, compromised mitigation programmes, weak implementation and a lack of accountability in pursuing outcomes and targets. Recognising these problems, he explores the limits and possible costs of relying more on adaptation as a response to climate change.

The next three chapters look at a number of issues surrounding the centrepiece of climate change policy – market-based carbon offset management. In two related chapters *Timothy Cadman* and *Tek Maraseni* apply the analytical governance framework outlined in this Introduction to two key policy instruments, the emergent REDD+ programme and the now well-established CDM. After a background discussion on the emergence, architecture and institutional expression of carbon offset mechanisms, *Cadman* finds that stakeholder perceptions of the governance of UNFCCC REDD+-related negotiations vary considerably among participants from the global North and South and across stakeholder sectors, with implications for the current design and future directions in market-based approaches to climate change management. *Maraseni* focuses on the CDM. He identifies two interrelated problems concerning the implementation capacity of the CDM to foster the behaviour change necessary for sustainable development and solve the problem of enhanced GHG emissions. CDM projects have not been able to capitalise on sustainable development objectives because of the distracting incorporation of hydrofluorocarbon projects within the Mechanism and problems associated with the so-called 'unilateral' CDM policy. In addition, despite the disproportionate level of attention, emissions of GHGs have been growing faster in developing countries than elsewhere. Those who have most extensively adopted CDM projects, namely China and India, show little slowdown in domestic emissions, and other developing countries have missed out on CDM investment. Given the importance of investment in such market-based environmental policy instruments, *Matthew Haigh* looks at climate finance and how financial institutions, such as pension funds and insurance companies, have interpreted and used UN-issued climate change management policies. While policymakers have been eager to appropriate the discourse of financial services, they are yet to supply guidance on how policies might best be applied to wealth portfolios. Financial institutions have been left to arrive at determinations without the basic architecture that usually accompanies their decision-making.

The next four chapters comment on four critical policy arenas outside the formal climate change space, but which are nevertheless affected by climate change and climate policies. *Jeff Gow* explores the impacts of climate change on health systems and the governance of service provision. He concludes that the risks to health of climate change will be incremental, increasing the burden of already occurring diseases like malaria and dengue fever and increasing the incidence of other major killers like diarrhoea and malnutrition. He contrasts the governance of the campaign against Human Immunodeficiency Virus Infection/Acquired Immunodeficiency Syndrome (HIV/AIDS) with health responses within the climate regime, which are fragmented, subject to special interests, uncoordinated and in dire need of integration. *Jamie Pittock* looks at water availability, which now afflicts a vast portion of the globe, and nearly a third of the world's people lack adequate access to water and sanitation. It is little appreciated that many policy responses to climate change are having perverse impacts on water resources and freshwater ecosystems. He finds little interplay between UNFCCC and other water-related environmental agreements and offers some pointers for governance reform to maximise the integration of climate change responses with sustainable water management. In the context of climate change and its impacts on food availability, *Nick Rose* provides a critical analysis of the differences between the 'food security' and the 'food sovereignty' approaches to the production, distribution and allocation of food resources and looks to alternative governance models to provide a solution to neo-liberal, market-driven systems. *Guilherme Lambais* and *Guilherme Gonçalves* challenge the legitimacy of contemporary energy governance in the context of climate change. They suggest instead a multi-level, multi-stakeholder approach based on renewable energy innovation and co-evolution of supply and demand, which they believe will kick-start renewable energy markets.

The last four chapters look at the ways in which a variety of societal actors are affected by, or influence, the institutional responses to the governance of climate change management. In two related chapters, *Richard Hil* and *Andrea Berringer* explore climate change and conflict and the effects on displaced persons. *Hil* describes the emerging relationships between climate change, conflict and displacement and provides two specific case studies: Papua New Guinea and Italy. In crafting policy responses to avoiding conflict, he stresses the need to acknowledge socio-cultural differences and an appreciation of the capacity of these differences to generate all manners of disruptions between and within nations. *Berringer* investigates climate change-induced migration and the ability of conventional refugee regimes to tackle this emerging problem. She provides a number of recommendations as to how to make sure that the needs of climate-related migrants are more effectively represented within these regimes into the future. *Heather Zeppel* looks at the role of local government in responding to climate change. She looks at the ICLEI and its CCP programme. She finds that an application of a governance analysis reveals who holds the power and influence with such networks. While there is much positive news in local and municipal action, the global climate agenda is dominated by developed countries, and greater effort is needed to integrate developing countries more meaningfully. *Julie Cotter* discusses how the NGO sector can exert pressure on GHG-emitting organisations by increasing the transparency of climate change management, while simultaneously shaping behavioural norms among major stakeholder groups. She looks at the development of the Climate Disclosure Standards Board (CDSB) as an NGO initiative to establish a comprehensive reporting framework for climate change information. Integration of this framework into mainstream corporate reporting represents a potentially significant improvement in institutional governance while also cementing the role of non-state interests in climate change management.

In the Conclusion, the contributors to this volume summarise their observations on the regimes investigated and provide a series of recommendations with a view to increasing the institutional legitimacy of the governance of the climate regime complex.

References

Andonova, B., Betsill, M. and Bulkeley, H. 2009. Transnational climate governance. *Global environmental politics*, 9 (2), 52–73.

Bäckstrand, K. and Lövbrand, E. 2007. Climate governance beyond 2012: competing discourses of green governmentality, ecological modernization and civic environmentalism. *In*: M. Pettinger, ed. *The social construction of climate change: power, knowledge, norms, discourses*. Aldershot and Burlington, VT: Ashgate, 123–148.

Barnett, J. 2010. Adapting to climate change: three key challenges for research and policy – an editorial essay. *Wiley interdisciplinary reviews: climate change*, 1 (3), 314–317.

Betsill, M.M. and Bulkeley, H. 2006. Cities and the multilevel governance of global climate change. *Global governance*, 12 (2), 141–159.

Betts, A. and Loescher, G. 2010. International cooperation in the refugee regime. *In*: A. Betts and G. Loescher, eds. *Refugees in international relations*. Oxford: Oxford University Press, 1–28.

Biermann, F. and Gupta, A. 2011. Accountability and legitimacy in earth system governance: a research framework. *Ecological economics*, 70 (11), 1856–1864.

Biermann, F., Betsill, M.M., Gupta, J., Kanie, N., Lebel, L., Liverman, D., Schroeder, H. and Siebenhüner, B. 2009. Earth system governance: people, places and the planet. *Science and implementation plan of the earth system governance project*. Earth System Governance Report 1, IHDP Report 20 Bonn, IHDP: The Earth System Governance Project.

Bulkeley, H. (2010). Climate policy and governance: an editorial essay. *Wiley Interdisciplinary Reviews: Climate Change*, 1 (3), 311–313.

Bumpus, A. and Cole, J. 2010. How can the current CDM deliver sustainable development? *Wiley interdisciplinary reviews: climate change*, 1 (4), 541–517.

Caballero-Anthony, M. 2010. Climate change and human security in south east asia: issues and challenges. *In:* A. Marquina, ed. *Global warming and climate change: prospects and policies in Asia and Europe*. London: Palgrave Macmillan, 393–413.

Cadman, T. 2011. *Quality and legitimacy of global governance: case lessons from forestry*. Basingstoke: Palgrave Macmillan.

Corbera, E. and Schroeder, H. 2011. Governing and implementing REDD+. *Environmental science and policy*, 14 (2), 89–99.

German Advisory Council on Global Change. 2008. *Climate change as a security risk*. London and Sterling, VA: Earthscan, http://wbgu.de/wbgu_jg2007_engl.pdf, accessed 13 January 2011.

Haas, Peter. 2002. UN conferences and constructivist governance of the environment. *Global governance*, 8 (1), 73–91.

Keohane, R. and Victor, D. 2010. The regime complex for climate change. Discussion paper prepared for The Harvard Project on International Climate Agreements, June 2010.

Kjaer, A. 2004. *Governance*. Cambridge and Malden, MA: Polity Press.

Koehn, P. 2010. Climate policy and action 'underneath' Kyoto and Copenhagen: China and the United States. *Wiley interdisciplinary reviews: climate change*, 1 (3), 405–417.

Koenig-Archibugi, M. 2006. Introduction: institutional diversity in global governance. *In:* M. Koenig-Archibugi and M. Zürn, eds. *New modes of governance in the global system: exploring publicness, delegation and inclusiveness*. Basingstoke: Palgrave Macmillan, 1–30.

Kooiman, J. 2000. Societal governance: levels, models, and orders of social-political interaction. *In:* J. Pierre, ed. *Debating governance: authority, steering and democracy*. Oxford: Oxford University Press.

Lammerts van Bueren, E. and Blom, E. 1997. *Hierarchical framework for the formulation of sustainable forest management standards.* Leiden: The Tropenbos Foundation.

Lederer, M. 2011. From CDM to REDD+ – what do we know for setting up effective and legitimate carbon governance? *Ecological economics,* 70 (11), 1900–1907.

Marquina, A. 2010a. Environmental challenges, conflict prevention and human security. *In:* A. Marquina, ed. *Global warming and climate change: prospects and policies in Asia and Europe.* London: Palgrave Macmillan, 481–497.

———. 2010b. From a socioeconomic approach to migration to the inclusion of environmentally induced migration in the Mediterranean. *In:* A. Marquina, ed. *Global warming and climate change: prospects and policies in Asia and Europe.* London: Palgrave Macmillan, 187–207.

Metz, B. and Cox, M. 2008. Integrating development and climate policies. *Climate Policy,* 8, 99–102.

Monreal, T. and Vargas Amelin, E. 2010. Effects of climate change on hydrological resources in Europe: the case of Spain. *In:* A. Marquina, ed. *Global warming and climate change: prospects and policies in Asia and Europe.* London: Palgrave Macmillan, 58–77.

Mulligan, S. 2006. The uses of legitimacy in international relations. *Millennium,* 34 (2), 349–375.

Mustafa, Maizatun. 2010. Water availability and policies in Asia. *In:* A. Marquina, ed. *Global warming and climate change: prospects and policies in Asia and Europe.* London: Palgrave Macmillan, 35–57.

Okereke, C. 2010. Climate justice and the international regime. *Wiley interdisciplinary reviews: climate change,* 1 (3), 462–474.

Paterson, M. and Stripple, J. 2007. Singing climate change into existence: on the territorialization of climate policy making. *In*: M. Pettinger, ed. *The social construction of climate change: power, knowledge, norms, discourses.* Aldershot and Burlington: Ashgate, 149–172.

Patz, J., Campbell-Lendrum, D., Holloway, T. and Foley, J. 2005. Impacts of regional climate change on health. *Nature,* 438 (17), 310–317.

Pierre, J. and Peters, B. 2000. *Governance, politics and the state.* Basingstoke: Palgrave Macmillan.

Sterling, Ana Yábar, 2010. IPCC Assessment reports: challenges presented. *In:* Antonio Marquina, ed. *Global warming and climate change: prospects and policies in Asia and Europe.* London: Palgrave Macmillan, 2010, 14–31.

Streck, C., et al. 2009. *Institutional options assessment developing an efficient, effective, and equitable institutional framework for REDD+ under the UNFCCC.* Washington, DC: Meridian Institute.

Thompson, M., Baruah, M. and Carr, E. 2011. Seeing REDD+ as a project of environmental governance. *Environmental science and policy,* 14 (2), 100–110.

Thynne, Ian. 2008. Symposium introduction. Climate change, governance and environmental services: institutional perspectives, issues and challenges. *Public administration and development,* 28, 327–339.

United Nations. (1992). Agenda 21: Programme of Action for Sustainable Development. United Nations Department of Public Information, New York.

1
The Discourses of Climate Change

Chris Taylor

Introduction

Climate change has been represented in a variety of ways. These representations have enacted their own discursive formations, which people discuss and act upon at local, national and global scales. Climate change was initially discussed within scientific disciplines and represented within a technical discourse. As it became popularised, through environmental organisations and the media, governments and intergovernmental bodies began to frame climate change within specific discursive formations, such as the United Nations Framework Convention on Climate Change (UNFCCC), the Intergovernmental Panel on Climate Change (IPCC) and the Kyoto Protocol (KP). These gave rise to forms of governance and discourse that have attained an almost hegemonic status, where climate change was framed within an overall neo-liberal governmental framework and network. As discursive formations of climate change were moved from science to government, they were transformed from a technical to a technocratic discourse. Institutional distance was created, resulting in the exclusion of other stakeholders and alternative discourses. Governance structures became elitist and exclusionary. The framing of climate change within global and national economic frameworks became the point of entry for stakeholders in climate change discussions. This chapter provides an overview of these discursive formations and provides suggestions for ways in which stakeholders can move past these obstacles so that greater participation can result in addressing environmental change at multiple spatial and temporal scales.

Climate change as a threat

> Climate change threatens the basic elements of life for people around the world – access to water, food, health, and use of land and the environment.
>
> (Stern 2006: 65)

Threat and insecurity have always been among the conditions of human existence. During pre-modern times, threats led to the assignment of blame. They were 'blows of fate' that assaulted human beings from outside and could be attributed to external gods, demons or nature (Beck 2006: 7). However, the threats that confront late modern society contrast those of the pre-modern era. As Beck (2006: 4) argues, these threats are the products of the successes of modern civilisation, as opposed to some external entity. These threats include pollution, resource degradation, scarcity and over-population. Of these threats, climate change has become the most prominent. Climate change is the threat posed to late modern society as a consequence of its own success. In the many influential publications on climate change, including those of the IPCC and the Stern Review, climate change is framed as a threat that is projected into the future. While changes have already been noted in the current global climate, these projected threats are portrayed as risks to 'basic elements of life for people around the world' (Stern 2006: 65). In this sense, climate change is not presented as a catastrophe located in the present but as an anticipation of a catastrophe, a risk. Beck elaborates on this notion of an anticipated catastrophe and risk:

> Risk is not synonymous with catastrophe. Risk means the anticipation of the catastrophe. Risks concern the possibility of future occurrences and developments; they make present a state of the world that does not (yet) exist. Whereas every catastrophe is spatially, temporally and socially determined, the anticipation of catastrophe lacks any spatio-temporal or social concreteness.
>
> (Beck 2007: 9)

Anticipated catastrophe is evident in Article 2 of the UNFCCC, whereby its ultimate objective is to achieve stabilisation of greenhouse gas (GHG) concentrations in the atmosphere at a level that would prevent 'dangerous anthropogenic interference with the climate system' (UNFCCC 1992). This has been equated with keeping global average temperature within a 2°C increase of the pre-industrial average. Excess of this threshold has introduced risks of global tipping points, where global ecosystems cease being sinks and become sources for GHG emissions. Hansen warns of these tipping points:

> The relevant scientists – those who know what they are talking about – realize that the climate system is on the verge of tipping points. If the world does not make a dramatic shift in energy policies over the next few years, we may well pass the point of no return.
>
> (2009: 171)

The impending catastrophes are the tipping points through which the world descends into dangerous climate change. This is where the change in the climate system, initially driven by an external forcing such as anthropogenic GHG emissions, is no longer required to sustain the new pattern of change (Russill and Nyssa 2009: 340). An example of this is the predicted loss of the Amazonian rainforest modelled under IS92a 'business as usual' GHG emission scenarios (Cox et al. 2004). Under the Hadley Centre climate–carbon cycle model, the authors predict local temperature increase of more than 10°K, along with a decrease in rainfall of up to 64 per cent. Such changes are predicted to result in the loss of 73 per cent vegetation carbon and 72 per cent soil carbon. Such a phenomenon is referred to as a feedback mechanism, whereby the ecosystem becomes a net source of GHG emissions, irrespective of whether anthropogenic GHG emissions cease. In this case, a tipping point has been exceeded.

The construction of this ecological (and social) catastrophe was modelled using a climate–carbon cycle general circulation model (GCM) (Cox et al. 2004: 139). This was based on the third Hadley Centre coupled ocean atmosphere model (Hadley Centre Coupled Model, version 3 – HadCM3) that was in turn coupled to an ocean–carbon cycle model (Hadley Centre Ocean Carbon Cycle Model – HadOCC) and a dynamic global vegetation model (Top-down Representation of Interactive Foliage and Flora Including Dynamics – TRIFFID). The models include variables such as movements of carbon within the ocean system, the state of the terrestrial biosphere in terms of soil carbon and the structure and coverage of five plant functional types. These models bring the predicted future collapse of the Amazonian rainforest into the present and represent it through graphs, tables and maps. Such representations make the impending catastrophe calculable.

The decline of the Amazonian rainforest is a side effect of the rapid modernisation of society, particularly in the global North. The modelling of it, along with other impending environmental catastrophes, constructs global risks and their distribution throughout space and time. They are part of modernity that reflects on itself, in what Beck (1992) has termed 'reflexive modernisation'. However, the reflexive does not necessarily mean more knowledge about risk, but more non-knowledge. This is not to be equated with denial but with the uncertainty of science in the face of risk (Beck 2007: 122). While modelling can represent a future risk into the present, it is, in essence, an 'abstraction of reality' (Demeritt and Wainwright 2005: 207). Climate models, no matter how sophisticated, can only provide a partial representation on a highly complex reality. They effectively reduce reality to an analytically simplified set of physical processes as a form of abstract reasoning (Demeritt 2001: 314).

Representing the threat as discourse

The representation of climate change as a threat cannot be reduced to any one particular discursive practice, formation or series of statements. As the discussion of its risk moved from scientific discussion into broader social and political debate, it was essentially transformed into a complex series of interpretations, and it generated a new network of social practices and relations. These are inherently complex and require multi-dimensional forms of critical analysis in order to examine their underlying social structures and respective relations of power and knowledge. Such revelations can provide critical insights into how new interpretations of the risks of climate change are created and how forms of governance emerge in the attempt to respond to those risks.

One particular form of analysis is through the discourses associated with the risks of climate change. There are numerous definitions of discourse. One of the most widely adopted is that of French philosopher Michel Foucault, where he describes discourse as a group of statements 'in so far as they belong to the same discursive formation' (1972: 131). Fairclough (2003: 124) expands on this definition by seeing discourse as a way of representing the world, including the processes, relations and structures of the material world; the mental world of thoughts, feelings and values; and the social world itself. In addition, discourses are also projective and representative of possible worlds that are different to the actual world. This can include modelled worlds, such as those that model risk. The relations between discourses also involve relations of power. Foucault elaborated on these relations, where power produces knowledge, and suggested that there is 'no power relation without the correlative constitution of a field of knowledge' (1977: 27).

Climate change represented as a change in a statistical average

In analysing 'climate change' as a term used in a variety of discourses, a good starting point is to analyse the discursive formation of the term 'climate'. It must be noted that the object of climate is not directly experienced. It is the weather that is experienced directly through the senses and its impacts that are identified in the natural and built environments. By contrast, climate is the abstract construction of averages over a defined period of time. For example, the IPCC (Solomon et al. 2007: 104) defines 'climate' as the 'average of weather' in the form of a statistical description. This average is determined for weather statistics measured over a 30-year period (Hulme et al. 2009: 199). In this sense, 'climate' is a statistical abstraction. It is a product of social practices and conventions that made it possible to construct a universal from multiple observed particulars (Demeritt 2001: 312). This quantification opened new interpretations and utilities. Comparative climatic analysis could be undertaken that relied on numerical data that provided for measurement and quantification. Climate was 'normalised' against these statistical baselines (Hulme et al. 2009: 198). It imposed order onto seemingly chaotic weather patterns.

A change in climate is a change in the long-term statistical average of weather. This essentially defines the term 'climate change'. According to the IPCC (Solomon et al. 2007: 104), this term refers to a change in the state of the climate that can be identified by changes in the mean and/or variability of its properties. These changes are required to last for an extended period of decades or more. However, the term 'climate change' differs from previous interpretations of past climate variations, because it is identified as the result of anthropogenic GHG emissions. This is encapsulated in the definition provided by the Framework Convention on Climate Change, which states that 'climate change' is a change in climate that can be attributed directly or indirectly to human activity (Solomon et al. 2007: 943).

Consensus in the scientific community points to the finding that human activities are modifying the concentration of atmospheric constituents that absorb radiant energy and that most observed warming over the past 50 years is attributed to the increase in GHG concentrations (Oreskes 2004: 1686). However, the risks of its threats and impacts are mostly represented in an abstract manner, resting on scientific models and calculations. These can be difficult to prove or refute on the basis of everyday experience (Beck 2007: 71).

Climate change and the discourse of limits to growth

> To achieve the ultimate objective of the Convention to stabilize greenhouse gas concentration in the atmosphere at a level that would prevent dangerous anthropogenic interference with the climate system, we shall, recognizing the scientific view that the increase in global temperature should be below 2°C, on the basis of equity and in the context of sustainable development, enhance our long-term cooperative action to combat climate change.
>
> (Copenhagen Accord 2009)

With the climate in a state of change, current political discussion has focused on an acceptable stabilisation level for a changed climate. The UN Conference of Parties (COP), European Union (EU) and other policy organisation proposals have selected a 2°C temperature target as the maximum, allowing warming to avoid 'dangerous anthropogenic interference with the climate system' (Randalls 2011: 598). While the target has been criticised by economists, scientists and politicians, it has re-introduced the concept of ecological limits into global environmental discourse.

Ecological limits gained international prominence in the early 1970s when environmental discourse was dominated by predictions of global ecological doom, accompanied with a priority on human survival (Hay 2002: 174–175). This was encapsulated in the Club of Rome's *The Limits to Growth*, whose authors warned of ecological limits being exceeded through accelerating industrialisation, rapid population growth, widespread malnutrition, depletion of non-renewable resources and a deteriorating environment (Meadows et al. 1972: 21). Its 'apocalyptic' message helped to explain why the message of environmental doom had such unprecedented impact on elite opinions (Hajer 1995: 81). These issues presented a global ecological crisis that was seen to threaten the existence of humanity.

In its overriding preoccupation with human survival, *The Limits to Growth* discourse sidelined 'new left' calls for freedom and citizen participation. In its place came discussions of resource rationing, increasing government intervention, centralisation and population control (Eckersley 1992: 13). It held the view that human populations were constrained by the operation of ecological laws, which were biological.

They were understood as having economic and political consequences and were expressed in the economic form of externalities impacting ecologically defined goods (Rutherford 1999: 53). This approach was anti-democratic and judged human nature harshly (Hay 2002: 174). However, ecological limits have been re-introduced into climate change discourses. A number of environment groups have been pushing for deep cuts to anthropocentric GHG emissions, with some advocating for 'zero emission' scenarios (see Beyond Zero Emissions). Plans to facilitate these scenarios have been introduced into public discussion, but as opposed to supporting *The Limits to Growth* discourse, they propose a new growth economy based on technologies free from emissions (Wright and Hearps 2010).

Climate change and the discourse of sustainable development

The Limits to Growth discourse was seen by many economic and political circles as unacceptable to the dominant paradigms of neo-liberalism and economic development, because it pitted environmental protection against development. This was evident in the largely failed attempt to achieve outcomes in the 1972 United Nations Conference on the Human Environment (UNCHE) in Stockholm (Bernstein 2001). Following this failure, environmental protection was seen to be achievable if it was nested within the norms of economic development. Early manifestations of this concept emerged during the early 1980s in the discursive formation of 'sustainable development'. The World Commission on Environment and Development (WCED) (1987) report, 'Our Common Future', often referred to as the 'Brundtland Report', formally defined 'sustainable development' as:

> [D]evelopment that meets the needs of the present without compromising the ability of future generations to meet their own needs.
> (1987: 43)

According to Bernstein (2001: 63), 'sustainable development' was the cornerstone of the WCED. It became the dominant conceptual framework for responses to international environmental problems and captured the imagination of world opinion (Bernstein 2000: 470). It proposed a new era of economic growth that was based on policies that 'sustain and expand the environmental resource base' (WCED 1987: 1). Integration of sustainability into development was one of the primary successes of the concept because it aimed to legitimate economic growth in the context of environmental protection. Accelerated economic growth was a top strategic priority. It encouraged a managed liberalism in the international economic order and integration of environmental protection goals into domestic development policies and international institutions (2002: 66). However, scientists were not the driving force behind the Brundtland Report, nor were they part of its consensus.

Climate change and the discourse of liberal environmentalism

The nesting of environmental protection within the paradigm of economic growth matured in the 1992 United Nations Conference on Environment and Development (UNCED). It expressed the integration of environmental protection into economic norms. Bernstein (2001) defines this as 'the compromise of liberal environmentalism'. Its norm complex supported the liberalisation of trade and finance as consistent with global environmental protection. He explains:

> It promotes sustained economic growth, free trade, privatisation of the commons and the use of market-based or other economic mechanisms [...] as the preferred means of environmental management.
> (2000a: 474)

This predicated international environmental protection on the promotion and maintenance of a liberal economic order. Since UNCED, trends have indicated increasing institutionalisation of liberal environmentalism (Bernstein 2000: 474). Trade liberalisation was perceived as having a positive impact on the environment by improving allocation of resources, promoting economic growth and increasing general welfare, provided that effective environmental policies were implemented (Bernstein 2001: 116).

Governing the threat through discourse

The governmentality of climate change

These discourses, including the construction of global risk, have made climate change an object of knowledge that can be governed or managed in specific ways (Rutherford 1999). An ensemble of institutions, descriptions, procedures, calculations and tactics has been formed to address the problems posed by climate change, resulting in specific and complex forms of power – in short, following Foucault, a governmentality (Foucault 2007; 2008). The activities of the state have increasingly taken the form of endeavours to shape, guide and direct conduct (Rose

1999: 3; Hindess 1996: 108). The most prominent and influential form of governmentality regarding the object of climate change is at the international level, where national governments around the world have formed institutional structures in which to address the object of climate change. Specific objectives that include the target of keeping the climate system within 2°C of pre-industrial average temperatures were sought. These include the UNFCCC, which has resulted in the COP. According to the FCCC, the COP is the 'supreme body' of the Convention. It consists of an association of all countries that are 'Parties' to the Convention. The purpose of the COP is to review the implementation of the Convention and examine the commitments of Parties in light of the Convention's objective, new scientific findings and experience gained in implementing climate change policies. It assesses the effects of the measures taken by Parties and the progress made in achieving the ultimate objective of the Convention.

According to Bodansky (2010: 31), there were two alternative models considered for the FCCC: (1) a general framework agreement on the 'law of the atmosphere', which was modelled on the 1982 UN Law of the Sea Convention, and (2) a convention specifically on climate change, which was modelled on the Vienna Ozone Convention. These would recognise the interdependence of atmospheric problems. The latter approach prevailed and took around three years to develop. The process began in December 1990, when the UN General Assembly established the Intergovernmental Negotiating Committee for a Framework Convention on Climate Change (INC/FCCC). Its purpose was to negotiate a convention for the 1992 UNCED Earth Summit at Rio. Bodansky (2001: 32) considers that two factors were critical. First, the 1992 UNCED deadline exerted substantial pressure on governments. Second, the desire for consensus decision-making gave individual countries substantial leverage over the final outcome. Bodansky claims that the discussions in the INC/FCCC followed a pattern common to international environmental negotiations. At first, little progress was apparent. The main negotiations materialised in the final months before UNCED, when governments realised that they would need to compromise if they wished to have a convention to sign at Rio. Its initial baseline was the 'framework agreement' model that was used in the preceding decade to address the acid rain and ozone issues. Signatories to the FCCC have been committed to providing a number of actions, which include national inventories of anthropogenic emissions by source and removals by sinks of all GHG emissions; publishing national and regional programmes containing measures to mitigate climate change; and promoting and cooperating in the development, application and diffusion of technologies, processes and practices that control, reduce or prevent anthropogenic GHG emissions (UNFCCC 1992).

From a governmentality perspective, the FCCC formed a specific technique or technology of government. Foucault referred to these technologies as the 'art of government', which consisted in ways of knowing an activity and how it may be carried out (Gordon 1991: 3). This means that governing requires specific practices for the production of truth and knowledge (Dean 1996: 28). It allows subjects, such as the FCCC, to divest itself of many of its obligations. Quasi-autonomous entities, such as signatories to the FCCC, could be governed at a distance by means of technologies of government. This has implemented a relationship of power–knowledge, where the techniques of power were developed on the basis of knowledge. In this sense, knowledge was generated through the national inventories of anthropogenic GHG emissions that then allowed for those emissions to become the targets for governing. Relations of power were exercised in that signatories were committed to meeting their respective targets. These targets represent techniques concerned with exercising power in the process of collecting knowledge (Fairclough 1992: 50). The broader suite of technologies used in the FCCC constitutes a means for the assemblage of knowledge. They were intended to achieve specific outcomes in the conduct of those who govern (Rose 1999: 52). These technologies shaped, normalised and instrumentalised conduct, thought, decisions and aspirations of signatories to achieve specific objectives through procedures of notation, collecting and presenting statistics. The knowledge generated from these procedures could then be transported to centres where calculations and judgements were made. These technologies enabled 'government at a distance', through indirect mechanisms of rule. Specific objects, such as national anthropogenic GHG emission profiles, were constructed in a particular conceptual form and made amenable to intervention and regulation (Rose and Miller 1992: 5). Thus, the climate change governmentality was constructed.

Ecological modernisation as a discourse of climate change

The Framework Convention on Climate Change and the COP all located actions in addressing the object of climate change within an overall paradigm of economic development. Article 3 of the FCCC stated its overall objective as the 'stabilisation of greenhouse gas concentrations in the atmosphere at a level that would prevent dangerous anthropogenic interference with the climate system'. However, this stabilisation had to be within a limit that allowed economic development to proceed in a 'sustainable manner'. As opposed to transforming the dominant paradigm of economic development, the FCCC's objective made it clear that it was economic development that was seen as providing the solution to address the threat of climate change. In this sense, climate change was seen as a failure in the market system that could be addressed through the imposition of a price on anthropogenic GHG emissions. By incorporating anthropogenic GHG emissions into the market system, the market was seen to have the capacity to correct this 'negative externality' of economic development.

This approach to addressing climate change was seen to fit within the norms of 'ecological modernisation'. Hajer (1995: 25) defines ecological modernisation in general terms as the discourse that 'recognises the structural character of the environmental problematique but nonetheless assumes that existing political, economic, and social institutions can internalise the care for the environment'. It introduces the concepts that make issues of environmental degradation calculable. It frames environmental problems through the combination of monetary units with discursive elements derived from the natural sciences. Environmental protection becomes a management problem that can be addressed by 'modernising' the economy. New techniques of environmental policymaking were introduced, such as the 'polluter pays principle', cost benefit analysis, the precautionary principle, tradable pollution rights and taxes on resources and emissions. Such developments provided a new role for science in environmental policymaking. Science was required to provide 'proof' of damaging effects that entangled it into policymaking discourse. The scientist was allocated the task of determining levels of pollution that the environment could be seen to endure. Importantly, ecological modernisation also shifted environmental protection away from the idea that it increased costs. Instead, efficiency in process and production was emphasised: nature was conceptualised as a public good in place of the idea that it was a free and unlimited resource (Hajer 1995: 26).

Oels (2005: 196) further defines ecological modernisation into two categories: 'weak' ecological modernisation and 'strong' modernisation. The weak variant believes that the free market context, combined with (limited) government incentives, will spur technological innovation that solves the ecological crisis in a cost-efficient manner. This has been the most common form of ecological modernisation in political practice and best matches the current paradigm of economic development within advanced liberal governments. However, the strong variant of ecological modernisation demands broad public participation in environmental decision-making, and it is more critical of current political and social institutions. The weak variant of ecological modernisation has been dominated by discourses of economic development. It has been much less informed by scientific discourses (Bernstein 2001).

Governing climate change through technocratic expertise

As climate change threats rose to prominence during the 1980s and 1990s, they became associated with a public policy tradition that accorded substantial status to scientific expertise harnessed by administrative structures. According to Dryzek (2006: 75), this nexus of science, professional administration and bureaucratic structure was used in many policy settings, such as natural resource management. He refers to this discourse as 'administrative rationalism'. This constructed a problem-solving discourse, taking the structural status quo of economic development and advanced liberalism as given. It had a strong conception of the nature of government, which was the administrative state. In this context, governing was not about democracy but about rational management by experts and managers in the service of a defined public interest (2006: 87). These experts are institutionally and politically centralised in their authority, which creates an institutional distance. This limits the 'space for dialogue, disagreement or differing points of view' (Iedema 1997: 74). 'Trained experts', who rule by virtue of their specialised knowledge, make the decisions. This gives rise to technocratic rationality that is linked to hierarchical administration (Eden 1999: 1297). According to Lemke (1995: 70), the technocratic elite claims a right to rule on the grounds of its ability to use expert knowledge to solve social problems.

One of the major problems with technocratic discourse is that it can be exclusionary. Critics of technocratic elites are portrayed as 'non-expert', 'non-rational' and unable to evaluate the contributions of scientific and business experts to environmental policy (Eden 1999: 1302). This can include scientists with expertise in the discipline of climatology, but whose findings and opinions fall outside the dominant paradigm of economic development. The problem inherent in this discourse is the notion of power and knowledge centralised at an apex. Agents at the apex are assumed to have greater knowledge than those at subordinate levels. They are justified in assigning tasks and coordinating operations. However, Dryzek (2006: 93) explains that complexities surrounding issues defy the scope of centralisation. It is impossible for knowledge about these complexities to be contained within a centralised authority.

Conclusions and recommendations

Transitioning to a participatory model of governance

Within the dominant paradigms of economic development, the action to address the threat of climate change has been to activate technologies of government that fall within traditional modernisation. These are abstract and technocratic. The institutional response has been to address the threat of climate change as a consequence of modernity with more modernity. According to Beck (2007), the modernisation of the threats caused by modernisation will create additional risks. These are an inevitable part of a risk society, a consequence in the radicalisation of modernity (Beck 1992). In this conception, aggregative democracy is the primary technology of government. Voting is the mechanism for selecting preferences and determining outcomes, in stark contrast to discursive democracy, in which dialogue and discussion are the primary tools (Cadman 2009: 37). Cadman's construction of governance seeks the development of an institution that is participative and deliberative. It tests for substantive and behavioural outcomes, thereby determining the legitimacy of governance quality (Cadman 2009: 100). In the previous discussion of the dominant discourses of climate change, only the strong variant of ecological modernisation could accommodate such technologies of government. Alternatively, the discourses of climate change can shift away from the dominant paradigm of economic development into one that places primacy on the well-being of environment (and people). However, this seems unlikely in the present context, given the hegemony of economic development in social and political spheres.

References

Beck, U. 2007. *World at risk.* Cambridge: Polity Press.

———. 2006. *The society of risk: towards a new modernity*. Barcelona: Paidos.

———. 1992. *Risk society: towards a new modernity.* London: SAGE Publications.

Bernstein, S. 2001. *The compromise of liberal environmentalism.* New York: Columbia University Press.

Bernstein, S. 2000. Ideas, social structure and the compromise of liberal environmentalism. *European journal of international relations,* 6 (4), 464–512.

Bodansky, D. 2010. The Copenhagen climate change conference: a postmortem. *The american journal of international law*, 104 (2), 230–240.

Cadman, T. 2009. *Quality, legitimacy and global governance: a comparative analysis of four forest institutions.* PhD Thesis, University of Tasmania.

Copenhagen Accord. 18 December 2009. *In*: UNFCCC, Report of the Conference of the Parties on Its Fifteenth Session [hereinafter COP Report and session number], Addendum, at 5, UN Doc. FCCC/CP/2009/11/Add.1 (30 March 2010) (advance version).

Cox, P.M., Betts, R.A., Collins, M., Harris, P.P., Huntingford, C. and Jones, C.D. 2004. Amazonian forest dieback under climate-carbon cycle projections for the 21st century. *Theoretical and applied climatology,* 78, 137–156.

Dean, M. 1996. Putting the technological into government. *History of the human sciences,* 9 (3), 47–68.

Demeritt, D. and Wainwright, J. 2005. Models, modelling and geography. *In:* N. Castree, A. Rodgers and D. Sherman, eds. *Questioning geography.* Oxford: Blackwell, 206–225.

Demeritt, D. 2001. The construction of global warming and the politics of science. *Annuals of the association of american geographers,* 91 (2), 307–337.

Dryzek, J. 2006. *The politics of the earth: environmental discourses.* Second edition. New York: Oxford University Press.

Eckersley, R. 1992. *Environmentalism and political theory: towards an ecocentric approach.* Albany: State University of New York Press.

Eden, S. 1999. 'We have the facts'—how business claims legitimacy in the environmental debate. *Environment and planning A,* 31 (7), 1295–1309.

Fairclough, N. 2003. *Analyzing discourse: textual analysis for social research.* New York: Routledge.

Fairclough, N. 1992. *Discourse and social change.* Cambridge: Polity Press.

Foucault, M. 2008. *The birth of biopolitics: lectures at the college de France 1978–79.* New York: Palgrave Macmillan.

Foucault, M. 2007. *Security, territory and population: lectures at the college de France 1977–78.* New York: Palgrave Macmillan.

Foucault, M. 1994. Technologies of the Self. *In:* P. Ranibow and N. Rose, eds. *The essential Foucault.* New York: The New Press, 145–169.

Foucault, M. 1977. *Discipline and punish: the birth of the prison.* Harmondsworth: Pengiun Books.

Foucault, M. 1972. *The archaeology of knowledge.* New York: Routledge.

Gordon, C. 1991. Governmental rationality: an introduction. *In:* G. Burchell, C. Gordon, and P. Miller, eds. *The Foucault effect: studies in governmentality.* Chicago: The University of Chicago Press.

Hay, P. 2002. *Main currents in western environmental thought.* Sydney: UNSW Press.

Hajer, M. A. 1995. *The Politics of environmental discourse: ecological modernisation and the policy process.* Oxford: Oxford University Press.

Hansen, J. 2009. *Storms of my grandchildren: the truth about the coming climate catastrophe and our last chance to save humanity.* New York: Bloomsbury.

Hindess, B. 1996. *Discourses of power: from Hobbes to Foucault.* Cambridge: Blackwell Publishers.

Hulme, M., Dessai, S., Lorenzoni, I. and Nelson, D.R. 2009. Unstable climates: exploring the statistical and social constructions of 'normal' climate. *Geoforum,* 40, 197–206.

Iedema, R. 1997. The language of administration: organizing human activity in formal institutions, *In:* F. Christie and J.R. Martin, eds. *Genre and institutions.* London: Continuum.

Lemke, J. 1995. *Textual politics: discourse and social dynamics*. London: Taylor and Francis.

Meadows, D.H., Meadows, D.L., Randers, J. and Behrens W.W., III. 1972. *The limits to growth: a report for the club of Rome's project on the predicament of mankind*. London: Earth Island.

Oels, A. 2005. Rendering climate change governable: from biopower to advanced liberal government? *Journal of environmental policy & planning*, 7 (3), 185–207.

Oreskes, N. 2004. The scientific consensus on climate change. *Science*, 306 (5702), 1686.

Randalls, S. 2011. History of the 2°C climate target. *WIREs climate change*, 1 (4), 598–605.

Rose, N. 1999. *Powers of freedom: reframing political thought*. Cambridge: Cambridge University Press.

Rose, N. and Miller, P. 1992. Political power beyond the state: problematics of government. *The British journal of sociology*, 43 (2), 173–205.

Russill, C. and Nyssa, Z. 2009. The tipping point trend in climate change communication. *Global environmental change*, 19 (3), 336–344.

Rutherford, P. 1999. The construction of environmental awareness. *In:* E. Darier, ed. *Discourses of the environment*. Malden: Blackwell Publishers.

Solomon, S., Qin, D., Manning, M., Chen, Z., Marquis, M., Averyt, K.B., Tignor, M. and Miller, H.L. (eds.) 2007. *Contribution of working group I to the fourth assessment report of the Intergovernmental Panel on Climate Change*. Cambridge: Cambridge University Press.

Stern, N. 2006, *The economics of climate change: the Stern review*, Cambridge: Cambridge University Press.

United Nations. (1992). Agenda 21: Programme of Action for Sustainable Development. United Nations Department of Public Information, New York.

Wright, M. and Hearps, P. 2010. *Australian sustainable energy, zero carbon Australia stationary energy plan*. Melbourne: Melbourne Energy Institute, University of Melbourne.

2
A Cooling Climate for Negotiations: Intergovernmentalism and Its Limits

Fred Gale

Introduction

Global negotiations to create international policy regimes to govern specific issue areas are conventionally conducted at intergovernmental forums. In the post-war period, and especially from the 1980s onwards, a wide range of such negotiations has occurred in economic, social and environmental policy arenas. Analysts have studied several success stories, including especially the 1986 moratorium on whaling introduced by the International Whaling Commission, the 1987 Montreal Protocol for phasing out of substances that deplete the ozone layer and the 1989 Basel Convention to restrict the trade in hazardous waste (Clapp 2002; French 1997; Peterson 1992).

In contrast to these timely and reasonably effective responses by states to international problems are those intergovernmental negotiations that fall into a second category and that have resulted in interminable discussions and produced weak and ineffective outcomes. The global forestry 'regime' is the paradigm case (Gale 1998; Humphreys 2006). The post-1992 negotiations on climate change also appear to fall into the second category of intergovernmental efforts. The optimism that greeted the signing of the United Nations Framework Convention on Climate Change (UNFCCC) at the United Nations Conference on Environment and Development (UNCED) and the negotiation of the 1997 Kyoto Protocol (KP) has given way to a renewed pessimism on the capacity of states to negotiate a timely, effective and equitable package. Notably, the failure of the United States to ratify the KP and subsequent acrimonious discussions between developed and developing countries over burden sharing have created an impasse. The ever-more urgent warnings by scientists at the Intergovernmental Panel on Climate Change (IPCC) on the effects of dangerous climate change thus appear to fall on the ever-deaf ears of state representatives who are forced to prioritise the economic platform of sustainability over its associated social and environmental platforms (IPCC 2007; Levy 2005).

The current incapacity of intergovernmentalism to deliver tough action on climate change is calling into question the output legitimacy of this form of global governance. Suspicion is growing, moreover, that the output legitimacy crisis of the UNFCCC and Kyoto arrangements is a consequence of significant deficiencies in their internal governance arrangements. In other words, these negotiating forums also lack input legitimacy. In particular, the marginalisation of scientists, environmentalists, indigenous peoples and affected populations in small island states means that greater weight is placed on protecting short-term economic and social interests to the detriment of the planet's ecological future (Dimitrov 2010b; Lynas 2010; McGregor 2011; Vormedal 2008).

This chapter applies elements of Cadman's 2011 framework to examine the governance quality of the UNFCCC and KP climate negotiations as they have evolved from the 1st Conference of the Parties (COP 1) in 1993 to the conclusion of COP 17 in Durban, South Africa, in December 2011. The analysis reveals serious deficiencies in the quality of intergovernmentalism for managing the climate change issue due to inadequate structures for meaningful participation and productive deliberation. It argues that these deficiencies have led to the current negotiating impasse and that environmental NGOs (ENGOs) are unable to significantly influence outcomes in consequence: rapid and effective change is more likely to be achieved by refocusing ENGO pressure directly on market actors than on attempts to control them via intergovernmental regulation.

Intergovernmentalism as global governance

Despite almost five decades of globalisation, states remain the dominant institution for negotiating international agreements. This reflects the fact that globalisation is largely an economically led phenomenon with significant lags in adjustment by political and social institutions. In the 1980s, there was considerable optimism about the capacity of states to negotiate global environmental regimes, fuelled by some major success stories with the ozone layer, hazardous waste trade and ocean dumping issue areas (O'Neill 2009). A sense prevailed at the time that the international community of states as represented by their governments could come together to negotiate timely, effective and binding agreements to successfully tackle environmental problems (French 1997). Despite these successes, however, a number of important environmental issues, most notably forestry and fisheries, appeared to defy the international regime logic, and even where agreements were reached, they failed to include binding targets to effectively tackle the issue or side payments to compensate those affected (Gale 1998; Gale and Haward 2011; Humphreys 2006). Climate change has emerged as the paradigmatic case of state failure. The unpalatable economic consequences of strong sustainability in a context of rising global business power have provoked a public/private producer backlash, resulting in weak interpretations of the concept of sustainable development, distorted indicators of its achievement and half-hearted efforts to monitor, report and verify (MRV) progress on the ground (Fuchs 2007; Levy 2005). The consequence has been ongoing high levels of greenhouse gas (GHG) pollution (UNEP 2012: 7–8).

In the early 1990s with the increasing recognition that government-to-government negotiations were compromising away the environment, ENGOs began to conceptualise new ways of aggregating, representing and mediating interests at the global level. Two broad alternative mechanisms emerged: multistakeholderism and chamberism (Gale, forthcoming). Multistakeholderism involves a central, usually self-appointed, governing body consulting with a wide range of stakeholders in the process of determining policy and practices and is epitomised by the Marine Stewardship Council (Gale and Haward 2011). Chamberism is a different model whereby anyone can join as a member and members are allocated to constituency-based chambers representing a range of interests, most commonly economic, social or environmental. This model is exemplified by the Forest Stewardship Council (FSC), and in this particular instance, chambers are further divided into two, with members indicating whether they come from rich developed countries or poor developing countries (Cadman 2011: 53–55; Gale and Haward 2011: 48–90). Despite the emergence of alternative, new mechanisms for governing global natural resource management issues, the dominant mechanism is intergovernmentalism, and it is this mechanism that has been at the heart of efforts to slow global warming and prevent dangerous climate change.

Climate change negotiations and intergovernmentalism

In 1990, the United Nations General Assembly established an Intergovernmental Negotiating Committee (INC) 'to negotiate a framework convention, containing appropriate commitments, and any related legal instruments as might be agreed upon' (quoted in Boisson de Chazournes 2008: 2). The objective, subsequently realised, was to agree upon a framework treaty for signature by delegates at the June 1992 UNCED conference in Rio de Janeiro. At UNCED, most governments signed the UNFCCC, which subsequently entered into force on 21 March 1994. Unusually in international law, the INC continued to meet to secure quick implementation of the agreement following its ratification. Six INC meetings were held between 1992 and 1995, with discussions centring on such issues as adequacy of commitments, rules of procedure, structure of the Bureau, operation of subsidiary bodies, methodologies for measuring emissions and financial support. With progress made on many of these issues, COP 1 was able to debate and reach agreement on the need to negotiate a Protocol to the treaty (Boisson de Chazournes 2008; *ENB* 1995).

The UNFCCC's overriding objective is the 'stabilization of greenhouse gas concentrations in the atmosphere at a level that would prevent dangerous anthropogenic interference with the climate system' (UNFCCC 1992, Article 2). The core principles by which this objective would be reached included recognition of the 'common but differentiated responsibilities' of developed and developing states, the need to give special consideration to those countries especially vulnerable to the effects of climate change as well as the precautionary, economic efficiency, development and free trade principles. To achieve the objective in accordance with these principles, developed states (Annex I countries, which included countries in economic transition like Russia) committed themselves to stabilising GHG emissions at 1990 levels by 2000 and to produce annual reports setting out their GHG inventories, and their mitigation and adaption policies. Annex II countries (a subset of Annex I countries composed of most Organisation for Economic Cooperation and Development (OECD) countries) also agreed 'to provide funding to cover full incremental costs of implementation of general commitments by developing countries and assist the particularly vulnerable States in meeting costs of adaptation' (Boisson de Chazournes 2008: 3).

While additional meetings of the INC enabled COP 1 to negotiate a range of matters quickly, including agreement on a Berlin Mandate to initiate negotiations on a Protocol to the Convention, a number of issues were unresolved. These included especially a resolution of the UNFCCC's rules of procedure. The Convention states that 'each of the parties to the Convention shall have one vote' and this applies to all Parties except 'regional economic integration organizations' like the European Union (EU). Only states can become Parties to the convention. With regard to regional integration organisations, they can 'exercise their right to vote with a number of votes equal to the number of their member States that are Parties to the Convention', although they can only do so if none of their Member States vote (UNFCCC 1992, Article 18). The UNFCCC provides that the COP will 'Agree upon and adopt, by consensus, rules of procedure and financial rules for itself and for any subsidiary bodies'. The only recourse to voting provided for the Convention relates to treaty amendments. Article 15 (3) states: 'If all efforts at consensus have been exhausted, and no agreement reached the amendment shall as a last resort be adopted by a three-fourths majority vote of the Parties present and voting at the meeting.' Any amendments passed by such a procedure are only binding on those voting for them and depositing instruments of acceptance, so even these limited voting arrangements do not really resolve deadlock (UNFCCC 1992, Article 15 (4 & 5)).

The UNFCCC also established two 'subsidiary bodies' and empowered the COP to establish others as required. The two subsidiary bodies are the Subsidiary Body for Scientific and Technological Advice (SBSTA) and the Subsidiary Body for Implementation (SBI). The purpose of the 'multidisciplinary' SBSTA is to provide 'timely information and advice on scientific and technological matters relating to the Convention'. By contrast, the purpose of the narrower SBI, composed of climate science experts, is to 'assist the Conference of the Parties in the assessment and review of the effective implementation of the Convention' (UNFCCC 1992, Articles 9 & 10). Both bodies are composed of government representatives who are 'competent in the relevant field of expertise' or 'experts on matters related to climate change' respectively (UNFCCC 1992, Articles 9 & 10). A schematic overview of the UNFCCC's institutional arrangements is set out in Figure 2.1.

At subsequent meetings of the COP, the SBSTA debated on a wide range of issues related to the impacts of climate change, the vulnerability of different regions and groups and potential adaptation measures. Work was done on emissions from deforestation and forest degradation in developing countries, on mechanisms to promote the development

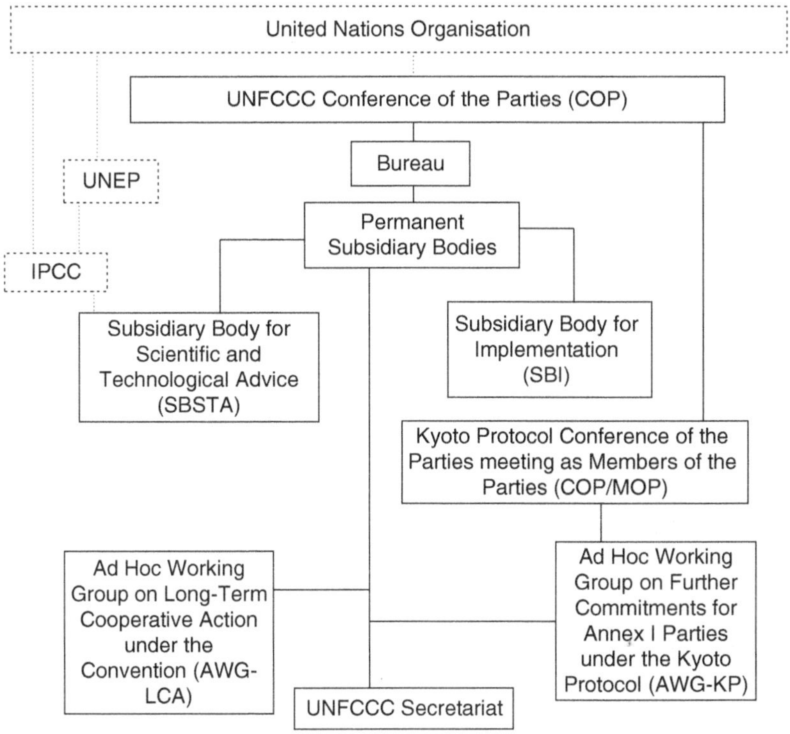

Figure 2.1 Basic institutional arrangements for UNFCCC climate change negotiations

Note: Lines indicate how different bodies relate to each other. Dotted lines indicate relationships external to the UNFCCC; solid lines indicate relationships internal to the UNFCCC. Bodies that operate under Convention rules are located in the centre and left-hand side of the diagram; bodies that operate under KP rules are located on the right-hand side. The UNFCCC Secretariat is located at the bottom of the diagram, signalling its coordination and secretarial role in servicing other UNFCCC bodies.

and transfer of environmentally appropriate technologies and on the preparation of guidelines for inventorying and measuring GHG emissions from Annex I Parties (UNFCCC 2012a). The purpose of the SBI, in contrast, was to monitor the implementation of the UNFCCC to determine if timetables and targets were being achieved. To perform this role, the SBI has scrutinised the information submitted by state parties in their national communications and emissions inventories (UNFCCC 2012b).

At the 1997 COP 3, Parties negotiated the KP, which set out more stringent expectations with respect to GHG reductions. In Annex A of the KP, six major GHGs along with key sectors and sources are listed for inclusion in a state's inventory of its GHG emissions. Annex B lists the actual commitments of each Annex I country, with most entries consisting of a reduction in GHG emissions over 1990 levels. A few countries, notably Iceland (110 per cent), Australia (108 per cent) and Norway (101 per cent) were permitted to increase emissions whereas most agreed to reduce them to 92 per cent (Kyoto Protocol 1998: Annex B). The KP establishes two ad hoc working groups: the Ad Hoc Working Group on Further Commitments for Annex I Parties under the KP (AWG-KP) and the Ad Hoc Working Group on Long-Term Cooperative Action under the Convention (AWG-LCA). The purpose of the AWG-KP is to negotiate the binding targets for the second commitment period after 2012 (UNFCCC 2012c). By contrast, the purpose of the AWG-LCA is to negotiate on the issue of non-Annex I contributions to reducing GHG emissions over time. The parcelling of these two issues into separate negotiating tracks created significant difficulties for subsequent negotiations, especially as developed countries increasingly targeted the need for binding commitments from developing countries.

Governance quality of the climate change negotiations

While the UNFCCC is an agreement made by and for governments, Article 7 (6) provides for the attendance of observers. The UN and its specialised agencies have an automatic right to attend as observers by virtue of their status as intergovernmental organisations (IGOs). Non-Member States also have a right to attend as observers. The agreement also extends conditional observer status to other bodies 'whether national or international, governmental or non-governmental' that are 'qualified in matters covered by the Convention' (UNFCCC 1992, Article 7(6)). The condition is that 'at least one third of the Parties present' do not object to the body seeking observer status.

Observers have been active participants at UNFCCC COPs. Cabré provides data on the attendance of non-governmental organisations (NGOs), Parties, UN Organisations, IGOs and Media at COPs 1–15 (2011: 11). Generally, the number of observers has increased over time with spikes at three major meetings: COP 3 in 1997 in Kyoto, COP 13 in 2007 in Bali and COP 15 in 2009 in Copenhagen. NGO attendance has increased from fewer than 1,000 at COP 1 to an average of just under 3,000 at subsequent meetings. Similarly, the number of IGOs attending COPs has increased from 23 at COP 1 to an average of almost 150 at subsequent meetings (ibid).

As a consequence of the high number of observers at COP meetings, in 2003 the UNFCCC Secretariat published a set of *Guidelines* for those attending meetings (UNFCCC Guidelines 2003). The *Guidelines* concern themselves mostly with ensuring that observers follow appropriate international etiquette such as displaying badges visibly, not harassing delegates, respecting UN flags and so forth (UNFCCC Guidelines 2003: 4). However, the *Guidelines* also draw the attention of observers to the Draft Rules of Procedure governing the UNFCCC set out in rules 7 and 30 of UNFCCC/CP/1996/2 and in Decision 18/CP.4. Rule 7, on the one hand, restates the right of observers to attend UNFCCC sessions unless one-third of the Parties decide otherwise. Rule 30, on the other hand, provides that COP meetings shall be held in public unless otherwise decided, while SBSTA and SBI meetings shall be held in private unless otherwise decided. Observers have benefitted from Decision 18/CP.4, which permits presiding officers 'to invite representatives of intergovernmental and non-governmental organizations to attend as observers any open-ended contact group established under the Convention process'. The *Guidelines* specifically note, however, that delegates should carefully observe whether these meetings are public or private. It states that 'In this context, meetings designated as CLOSED are not open to observers' (UNFCCC Guidelines 2003: 5, capitals in original).

Subsequent negotiations with respect to the 1997 KP, the 2007 Bali Mandate (designed to lead to a new commitment period under tougher, binding targets) and the 2009 Copenhagen Climate Conference (COP 15) have all been conducted under the above arrangements. These arrangements not only formally provide for negotiations to be conducted by states party to the UNFCCC but also permit a large number of non-state actors, including the United Nations, IGOs and NGOs, to attend meetings and participate in public deliberations on important matters related to climate change (Betsill and Correll 2008).

The potential for non-state actors, especially ENGOs, to utilise these organisational structures to influence outcomes has been much commented on in the literature (for example Betsill 2002; Betsill and Correll 2008; Cabré 2011; Giorgetti 1999; McGregor 2011). Betsill claims that members of the major environmental coalition attending the UNFCCC, the Climate Action Network (CAN), 'directly shaped the nature of debates around emissions trading and sinks, and indirectly influenced negotiations on targets and timetables by putting pressure on delegates from the European Union and the United States' (2008: 65). Using a similar framework, but examining business organisations, Vormedal concludes that Business International NGO (BINGO) activities showed that 'BINGOs appear to have exercised considerable influence on the process of negotiating a regulatory design for including CCS [Carbon Capture and Storage] as a CDM project activity' and that his case study 'also established BINGO goal attainment vis-à-vis the African delegations, which changed their position from opposition to support for CCS in the CDM from the Nairobi to the Bali meeting' (Vormedal 2008: 62).

However, there are strong grounds for viewing this agency-centred perspective as overemphasising the capacity of business and civil society organisations to influence international negotiations. The ways institutions structure interest representation and manage negotiations are highly significant determinants of outcomes. External actors are significantly disadvantaged if they do not have a seat at the decision-making table (Young 2000: 8–13). From this perspective, the capacity of organisational outsiders to significantly influence internal negotiations and outcomes is much less. The Copenhagen Climate Summit negotiations offer a dramatic example of just how peripheral the actions of outsiders can be when state delegates raise the stakes and engage in a form of climate power politics.

The difficulty non-state actors confront is this: the UNFCCC is an IGO and its basic mode of interest representation is defined by the rules of multilateralism. In this system of governance, state representatives are the only officially recognised actors, and representatives of non-state actors are only allowed to be present when state delegates permit them to be. In many international negotiations, including those occurring within the purview of the UNFCCC, non-state actors are permitted to have influence up to the point when tough trade-offs need to be made, at which point they are excluded, along with a substantial number of state actors (Dimitrov 2010b; Schott and Watal 2000). This process was formalised in the infamous World Trade Organisation (WTO) 'Green Room' discussions that were an integral part of the trade negotiations up until developing countries objected to them at the aborted 1999 Millennium Round talks held in Seattle, Washington (Schott and Watal 2000). It has a substantial history in climate change negotiations too and was used extensively in the Copenhagen Climate Summit in an attempt to broker an eleventh-hour deal (Dimitrov 2010b; *ENB* 2009; Lynas 2009; McGregor 2011).

Popular and academic accounts of the Copenhagen Climate Summit paint a picture of an increasingly smaller group of states attempting to do a deal on climate by bargaining away core elements of an approach that would effectively address the issue (Dimitrov 2010b; *ENB* 2009; Lynas 2009). Perhaps the most authoritative account of what happened comes from Dimitrov (2010b), who not only is a student of global environmental governance but has also participated in the negotiations as a member of one of the European delegations. His account is a dramatic rebuttal of those who argue that international organisations are venues of potential influence for non-state actors. According to Dimitrov, the major actors who engaged in the negotiations were state coalitions, of which five were most salient. One was the EU, which sought significant progress on climate change, including a commitment to keep global warming below 2°C, binding emissions reduction targets for Annex I countries to secure a reduction in GHG emissions to 30 per cent below 1990 levels by 2020 and by up to 95 per cent by 2050, a non-binding commitment by developing countries to reduce GHG emissions to 30 per cent from currently predicted growth rates and a substantial Green Climate Fund (Dimitrov 2010b: 803). The EU could count on support from another group, the Alliance of Small Island States (AOSIS), which viewed their very survival as threatened by global warming and the anticipated rise in sea levels. AOSIS members made impassioned pleas for even tougher action, arguing their position from a high moral ground. Pragmatically, however, they recognised they were economically, militarily and politically weak and required alliances with other states if they were to influence outcomes (Dimitrov 2010b).

The nascent EU and AOSIS groups confronted a range of state-based coalitions that sought a substantially weaker outcome. Dimitrov identifies the Umbrella Group, the G-77/China and the BASIC group, as well as individual countries within each group, as opposing a substantive agreement. The Umbrella Group consisted of the former JUSCANZ countries (Japan, the United States, Canada, Australia and New Zealand), joined by Kazakhstan, Norway, the Russian Federation and Ukraine. According to Dimitrov, this coalition agreed with the 2°C objective, envisaged a 50 per cent reduction in GHG emissions by 2050, but sought to include all countries – especially the large developing country emitters of India and China – in any deal. The latter objective was strongly opposed by the developing country group, the G-77/China collectively and, when that began to disintegrate, by the BASIC coalition of Brazil, South Africa, India and China. In their opening statement to the Copenhagen Climate Summit, Sudan on behalf of the G-77/China stated that developing countries would oppose binding emissions targets for non-Annex I countries as they transgressed the previously accepted cornerstone principle of common, but differentiated, responsibilities (quoted in Dimitrov 2010b: 804).

According to Dimitrov, in the run up to Copenhagen, most delegates accepted that it would not be possible to negotiate a final, legally binding successor agreement to the KP. Progress in the two years since COP 13 had delivered the Bali Mandate had been too slow. Thus, delegates broadly aimed to reach agreement on a 'core decision' that would capture as much 'policy substance' as it could and pave the way for subsequent negotiations in 2010 (Dimitrov 2010b: 806). The debate was thus not over the desirability of a core decision but over the policy substance it would contain.

To negotiate the core decision, state representatives met in a variety of venues. Public statements were made in plenary meetings of the Conference of the Parties (UNFCCC COP) and the Conference of the Parties meeting as Members of the Parties (Kyoto COP/MOP). These formal sessions were open to the large number of business and civil society organisations until the arrival of the heads of state in the second week. However, negotiations on specific issues occurred in a wide range of committees, contact groups and informal meetings. There were meetings of the UNFCCC's subsidiary bodies, the SBSTA and SBI, dealing with technical issues related to Land Use, Land Use Change and Forestry (LULUCF) on the one hand and MRV on the other. Contact groups, 'friends of the Chair' and informal meetings, some open but many closed to observers, occurred with respect to these subsidiary bodies as well as with the KP's two Ad Hoc Working Groups: the AWG-LCA and the AWG-KP (Dimitrov 2010a; *ECO Newsletter* 2009; *ENB* 2009).

In most cases, observers were excluded from these meetings and especially from the more sensitive negotiations. Indeed, as the negotiations continued into the second week, and due to concerns over security, the Danish and UN authorities radically reduced business and civil society participation to 300, excluding the remaining observers, which were estimated to number over 25,000 (McGregor 2011). *ECO*, the environmental newsletter, expressed outrage at the lockout, noting that it could be a breach of the Aarhus Convention that expressly requires civil society participation in climate change negotiations. *ECO* noted: 'As this is a UN convention that came into force in Aarhus [Denmark], it makes Denmark's efforts to exclude civil society participants from this crucial conference unforgivable' (*ECO Newsletter* 2009: 4).

Many state delegations were subsequently sidelined when a Friends of the Chair group was established to progress negotiations. Initial opposition by the G-77/China to the Friends of the Chair proposal meant that negotiations became bogged down in procedural wrangling, resulting in little progress on the key issues that required resolution. Eventually, non-transparent negotiations resulted in a group of 25 states being nominated to the Friends of the Chair, but this only occurred on the last day of the meeting (Dimitrov 2010b: 808–809).

A further insight into what was actually going on in the Friends of the Chair group is provided by Mark Lynas (2009). Lynas, who was present during the Friends of the Chair discussions, makes the following observations.

> Here's what actually went on late last Friday night, as heads of state from two dozen countries met behind closed doors. Obama was at the table for several hours, sitting between Gordon Brown and the Ethiopian prime minister, Meles Zenawi. The Danish prime minister chaired, and on his right sat Ban Ki-moon, secretary-general of the UN. Probably only about 50 or 60 people, including the heads of state, were in the room. I was attached to one of the delegations, whose head of state was also present for most of the time.
>
> What I saw was profoundly shocking. The Chinese premier, Wen Jiabao, did not deign to attend the meeting personally, instead sending a second-tier official in the country's foreign ministry to sit opposite Obama himself. The diplomatic snub was obvious and brutal, as was the practical implication: several times during the session, the world's most powerful heads of state were forced to wait around as the Chinese delegate went off to make telephone calls to his 'superiors'.

In short, climate negotiations in Copenhagen came down to a small group of individuals representing states whose interests ranged from outright opposition (Saudi Arabia), to minimal ambition (the United States, the Russian Federation, China, India, Brazil), to modest ambition (the United Kingdom, Spain, Denmark, Germany, France), to strong ambition (Maldives).

In the early hours of the Saturday morning of the 19th December, delegates attempted to adopt the output of these Friends of the Chair negotiations in the form of the three-page Copenhagen Climate Accord. These talks resulted in an impasse between a relatively large number of delegations that thought it worthwhile to endorse the compromise and a small number of countries that were especially outraged at the process used to arrive at the agreement. A number of incendiary statements were made at this point, including one by the Sudanese government representative who argued that the values that underpinned the agreement were 'fascist' in nature because it would turn Africa 'into a furnace' (Dimitrov 2010a: 18). Cuba objected on procedural grounds, arguing that the 'group of representative leaders' was a 'gross violation of sovereign equality that is enshrined in the Charter of the United Nations and a mechanism that would impose decisions on the international community' (Cuban delegation quoted in Dimitrov 2010b: 811).

While the post-Kyoto and post-Bali negotiations were populated by a large number of official delegations and business and civil society observers provided ideas and input into a global climate treaty, much of this work was abandoned in the final days of negotiations as 'power politics' emerged as the dominant negotiating mode. A stand-off emerged between the United States on the one hand and China, India and Brazil on the other over all core elements of an agreement. None of the Parties was prepared to sign up to tough, binding quantified emission limitation and reduction objectives. The Obama administration was conscious of the fact that it could not get any legislation through Congress unless there were meaningful commitments from China and India, and China and India appeared to have decided that under no circumstances would they cede the basic principle underpinning Kyoto – that developed countries should cut their emissions first and that developing countries would only do so later based on voluntary targets that would not be subject to MRV (Dimitrov 2010b; *ENB* 2009).

Conclusions and recommendations

The conclusion to be drawn from the above analysis is that the governance quality of the UNFCCC is poor. Accounts of the Copenhagen Climate Conference highlight how the power politics of a few countries hijacked the negotiations and undid years of wider if not deeper intergovernmental cooperation. Both interest representation and decision-making were compromised by the UNFCCC's intergovernmental system of bargaining. Not only were social and environmental interests marginalised through the intergovernmental negotiating process, but so too were the interests of smaller states, including those grouped in AOSIS, and the views of scientists, as the negotiations moved towards their conclusion and the most important trade-offs were being made. Rather than securing a balanced representation of interests across the three platforms of sustainable development – economic, social and environmental – the negotiations ultimately brought together the leaders of powerful geographical territories who were legally bound to protect the 'national interest' of their peoples defined largely in economic terms.

The policy implications of this analysis are thus clear. ENGOs should abandon their efforts to influence global political negotiations on climate change and seek direct influence in global and domestic markets instead. This lesson, learned before with respect to tropical deforestation and fisheries depletion, is now evident with regard to climate change (Gale 1998; Gale and Haward 2011). In Canada, companies involved in the Alberta tar sands production chain should be a specific focus given the carbon intensity and unsustainability of this development. Since the United States is a major importer of tar sands oil, US ENGOs should also intensify their efforts to disrupt this trade. In Australia and China, the focus should instead be on the coal production commodity chain. China's booming but environmentally unsustainable industrial development strategy is currently based largely on domestic and imported coal, with Australia a major supplier.

In addition to efforts targeted at disrupting unsustainable production chains that are compromising the earth's atmospheric system, ENGOs should also foster alternative clean energy production chains linked to wind, water and solar power. In doing so, they need to pay attention to the full range of issues involved in their development. While alternative energy systems generate considerable environmental benefits, they can sometimes be economically costly and socially disruptive depending on how they are implemented. In short, ENGOs too must pay attention to governance quality and secure high levels of representation across the three platforms of sustainability as well as good processes that deliver deep deliberation across these platforms if the shift required to a clean energy future is to be achieved.

References

Betsill, M. 2008. Environmental NGOs and the Kyoto Protocol negotiations: 1995 to 1997. *In:* M. Betsill and E. Correll, eds. *NGO diplomacy: the influence of nongovernmental organizations in international environmental negotiations*. Cambridge, MA: The MIT Press, 43–66.

Betsill, M. 2002. Environmental NGOs meet the sovereign state: the Kyoto Protocol negotiations on global climate change. *Colorado journal of international environmental law and policy*, 13 (1), 49–64.

Betsill, M. and Correll, E. 2008. *NGO diplomacy: the influence of nongovernmental organizations in international environmental negotiations*. Cambridge, MA: The MIT Press.

Boisson de Chazournes, L. 2008. United Nations Framework Convention on Climate Change. *United Nations Audiovisual Library of International Law*, www.un.org/law/avl, accessed 24 March 2012.

Cabré, M. 2011. Issue linkages to climate change measured through NGO participation in UNFCCC. *Global environmental politics*, 11 (3), 10–22.

Cadman, T. 2011. *Quality and legitimacy of global governance: case lessons from forestry*. Basingstoke: Palgrave Macmillan.

Clapp, J. 2002. Seeping through the regulatory cracks. *SAIS review*, 22 (1), 141–155.

Clapp, Jennifer. 2002. The distancing of waste. *In:* T. Princen, M. Maniates and K. Conca, eds. *Confronting consumption*. Boston, MA: The MIT Press, 155–176.

Dimitrov, R.S. 2010a. Inside Copenhagen: the state of climate governance. *Global environmental politics*, 10 (2), 18–24.

Dimitrov, R. 2010b. Inside UN climate change conference: the Copenhagen conference.' *Review of policy research*, 27 (6), 795–821.

ECO Newsletter. 2009. *Copenhagen climate negotiations*, Newsletters 1 to 11, http://www.climatenetwork.org/eco-newsletters?field_event_nid=260, accessed 29 March 2012.

ENB (*Earth Negotiations Bulletin*). 2009. Summary of the Copenhagen climate change conference: 7–19 December 2009. *ENB*, 12 (459), 1–2.

———. 1995. Summary of the first Conference of the Parties for the Framework Convention on Climate Change: 28 March 1995 – 7 April 1995. *ENB*, 12 (21), 1.

French, H. 1997. Learning from the ozone experience. *In:* L. Brown, et al., eds. *State of the world report 1997*. Washington DC: The Worldwatch Institute, 151–172.

Fuchs, D. 2007. *Business power in global governance*. Boulder, CO: Lynne Rienner.

Gale, F. Forthcoming. Three models of interest mediation in global environmental governance: intergovernmentalism, multistakeholderism, chamberism.

Gale, F. 1998. *The tropical timber trade regime*. Basingstoke: Palgrave Macmillan.

Gale, F. and Haward, M. 2011. *Global commodity governance: state responses to sustainable forest and fisheries certification*. Basingstoke: Palgrave Macmillan.

Giorgetti, C. 1999. From Rio to Kyoto: A study of the involvement of non-governmental organizations in the negotiations on climate change. *New york university environmental law journal*, 7, 201.

Humphreys, D. 2006. *Logjam: deforestation and the crisis of global governance*. London: Earthscan.

IPCC (Intergovernmental Panel on Climate Change). 2007. *Climate change 2007: synthesis report: summary for policy makers*. Geneva, Switzerland: IPCC.

Levy, D. 2005. Business and the evolution of the climate regime: the dynamics of corporate strategies. *In:* D. Levy and P. Newell, eds. *The business of global environmental governance*. Boston, MA: The MIT Press, 73–104.

Lynas, M. 2009. How do I know China wrecked the Copenhagen deal? I was in the room. *The Guardian*, 22 December 2009, http://www.guardian.co.uk/environment/2009/dec/22/copenhagen-climate-change-mark-lynas, accessed 11 May 2012.

Lynas, M. 2010. High tide: how climate crisis is engulfing our planet. *Harper perennial*.

McGregor, I. 2011. Disenfranchisement of countries and civil society at COP 15 in Copenhagen. *Global environmental politics*, 11 (1), 1–7.

O'Neill, K. 2009. *The environment and international relations*. Cambridge: Cambridge University Press.

Peterson, M. 1992. Whalers, cetologists, environmentalists, and the international management of whaling. *International organization*, 46 (1), 147–186.
Schott, J. and Watal, J. 2000. Decision making in the WTO. *In:* Jeffrey Schott, ed. *The WTO after Seattle*. Washington, DC: The Peterson Institute, 283–301.
United Nations. 1998. *Kyoto Protocol to the United Nations Framework Convention on Climate Change*. Geneva, Switzerland: UNFCCC.
UNEP. 2012. *Global environmental outlook, GEO 5: summary for policy makers*. Nairobi: United Nations Environment Program.
UNFCCC (United Nations Framework Convention on Climate Change). 2012a. *Subsidiary Body for Scientific and Technological Advice: what is the SBSTA?* http://unfccc.int/bodies/body/6399.php, accessed 30 March 2012.
———. 2012b. *Subsidiary Body for Implementation: what is the SBI?* http://unfccc.int/bodies/body/6406.php, accessed 30 March 2012.
———. 2012c. *Ad Hoc Working Group on Further Commitments for Annex I Parties under the Kyoto Protocol: what is the AWG-KP?* http://unfccc.int/bodies/body/6409.php, accessed 30 March 2012.
UNFCCC (United Nations Framework Convention on Climate Change). 1992. http://treaties.un.org/doc/Treaties/1994/03/19940321%2004-56%20AM/Ch_XXVII_07p.pdf, accessed 24 March 2012.
UNFCCC Guidelines. 2003. *Guidelines for the participation of representatives of non-governmental organizations at meetings of the bodies of the United Nations Framework Convention on Climate Change*. March. Bonn, Germany: UNFCCC Secretariat.
Vormedal, I. 2008. The influence of business and industry NGOs in the negotiation of the Kyoto mechanisms: the case of carbon capture and storage in the CDM. *Global environmental politics*, 8 (4), 36–65.
Young, I. 2000. *Inclusion and democracy*. Oxford: Oxford University Press.

3

Gender and Climate Change: Stakeholder Participation and Conceptual Currency in the Climate Negotiations Regime

Lauren E. Eastwood

Introduction

Recent policy deliberations have been marked by an increase in advocacy for mainstreaming social justice concerns into institutions related to global environmental governance. Particularly in light of the well-documented gender and environment nexus, arguments in favour of mainstreaming gender into climate policy have gained considerable currency, resulting in some institutional changes in terms of how civil society participates in climate-related policy negotiations. This chapter addresses the terrain on which arguments for greater inclusion are made and highlights specific policy discussions that illustrate both the opportunities and the challenges for stakeholders who are interested in shaping policy outcomes.

As a result of the recognition of the gender/environment nexus (see, for example, Dankleman 2002, 2010) and the intensive social movement organising around women's concerns over the past few decades, gender has become integrated into the language of international environment and development policy. However, policymaking – particularly at the international level – is notoriously ambiguous and abstract. Those involved in policy processes know that the language of policy tends to obscure on-the-ground realities. Terms that at first glance appear relatively clear (such as sustainable development) can ultimately be defined in myriad ways, depending on the context in which they are taken up (Escobar 1995; Eastwood 2005; Harvey 1996; Visvanathan 1991). The ambiguity of terms used in policymaking serves the purpose of allowing a variety of actors with varying interests to agree upon a text or policy outcome. Within specific governance regimes, particular terms become conceptually salient over time. This chapter analyses the ways in which the term gender has been used within the climate policy regime as a tool to advocate for particular policy outcomes that are inclusive of civil society stakeholders and potentially marginalised individuals. Significantly, the concept of gender has come to hold considerable currency within the climate regime. As such, it can be used strategically to attempt to influence particular policy outcomes that address social justice concerns. Additionally, as a result of the particular ways in which the meaning of the term gender is interpreted within the climate regime, opportunities exist for stakeholders to utilise the conceptual currency to ensure that policy outcomes address the importance of a variety of ecosystem services in addition to carbon sequestration.

Due to the highly politicised nature of climate negotiations, the strategic use of language in policymaking related to the climate regime merits attention. Cadman (Introduction) argues that climate change is one of the most significant and normatively embedded post-Rio multilateral regimes. Given the increasing prevalence of climate discussions in virtually all UN-based environmental and social policy processes, the normatively embedded dynamics associated with climate negotiations deserve elucidation. Actors engaged in the Rio +20 meetings in Brazil in 2012 are not unaware of the increasing politicisation of climate dynamics that includes a proliferation of climate concerns into UN policymaking processes that were originally designed to address other environmental or social issues. This dynamic indicates a further entrenchment of climate-related discourse – an increased embedding of the climate regime.

Even as the discursive terrain of climate negotiations has shifted as the regime evolves, actors engaged in climate policy have deployed the conceptual currency of such notions as gender to attempt to shape policy outcomes. As climate governance becomes more pervasive and expands to incorporate other policy processes (see Chapter 9), stakeholders have an immediate and increasing interest in shaping the policy processes so that the institutionalisation of the regime can serve to address their concerns. A central argument of this chapter is that gender is currently being used as one of the tools to address some of the concerns of stakeholders engaged in the climate policy negotiations processes. Specifically, as policy processes become more institutionalised, concerns arise as to whether these emerging institutions will prioritise carbon sequestration over other ecosystem and social benefits associated with forests and agriculture, thus potentially exacerbating global inequalities related to access to resources. Gale's contribution to this volume highlights the intransigence of United Nations Framework Convention on Climate Change (UNFCCC) negotiators in terms of being willing to address greenhouse gas (GHG) emissions. Instead, as powerful governments' participation in the UNFCCC negotiations has precluded policy moves that would limit GHG emissions significantly, the policy realm has expressed some political will regarding carbon sequestration as an alternative. It is in this context that gender provides important openings for stakeholders who engage in multilateral negotiations regarding climate change, as a means of pushing policy negotiators to adopt a more holistic approach to climate policy.

Conceptual currency and the discourse of gender at the UN

The conceptual currency associated with gender, and, in particular, with women's marginalisation from resources, is taken up by actors as a strategic mechanism to influence policy both discursively and practically. Dorothy E. Smith's understanding of discourse is instrumental here. As Smith argues, discourse can be specified to mean '... those forms of communication and interrelation that are mediated by texts.... These are distinctive forms of contemporary organization which intersect with the largely hierarchical structures of state, business, and other administered formal organizations' (Smith 1987: 214). Perhaps more importantly, Smith goes on to argue that '[t]heir ideological processes serve to coordinate sites of the ruling apparatus coming under different jurisdictions' (ibid). A key element of Smith's understanding for this analysis here is not only the way in which discourses in current society are textually mediated but also that they serve a coordinating function. They are not merely abstract ideological formations but instead serve to organise individuals' activities in disparate locations. Thus, the currency of concepts such as gender and stakeholder participation can be deployed in a variety of negotiations in disparate locations. In many ways, as their legitimacy as concepts becomes institutionalised, they can be taken up in a variety of policy arenas. While they transform over time, Arturo Escobar articulates, discourses do not replace each other completely but build upon each other as layers that can be only partly separated (1995: 195). Thus, while discursive formations shift over time, there is a history that is brought forward into the transformed meanings. Thus, while discourse speaks to a highly complex entity, it also speaks to a shared-in-common (albeit nebulous or ambiguous) meaning – a framework that can organise participants' strategies within policymaking arenas. David Harvey argues that discourses internalise a whole range of contradictory impulses and conflictual ideas derived from all of the other moments in the social process (1996: 174). Furthermore, due to the fact that they are abstract yet refer to politicised realities, the discourses themselves conceal a concrete political agenda in the midst of highly abstract, universalising and frequently intense moral argumentation (ibid).

What this speaks to is the importance and complexity of language in policymaking processes, where actors deliberately use structures of interest representation as well as notions of accountability and transparency to gain desired outcomes. The focus here is on the negotiations under the UNFCCC. However, actors engaged in the UNFCCC process understand the need to engage in multiple fora, as agreed-upon language can be taken up in multiple processes. In terms of gender, a key resource for stakeholders has been the negotiations that explicitly deal with human rights. Participants in the UNFCCC negotiations who are interested in mainstreaming gender into various levels of the policy texts have drawn upon policy processes that explicitly address dynamics associated with social equity. In terms of mainstreaming gender concerns into global climate policy processes, Lambrou and Pianna argue that it is only since the UNFCCC COP 8 in New Delhi in 2002 that gender has explicitly been addressed in the policy documents associated with the UNFCCC negotiations. And, according to their analysis, this treatment has been tangential (Lambrou and Pianna 2005: 1). Significantly, related to the manner in which interest groups are recognised within the UNFCCC process, a statement on behalf of women was given for the first time at COP 11 in Montreal. Proponents of mainstreaming gender into climate policy argue that mistakes in climate policy may lead to consolidating or even extending inequalities between men and women (Rohr 2009: 4). This is a key line of argument, as it becomes clear that gender becomes one of the ways to tie social concerns into texts that were originally developed primarily to address climate change as an environmental or physical phenomenon.

The disconnect between the ways in which climate was being conceptualised by policy makers and the concerns of a variety of stakeholders became apparent with the initial discussions within the UNFCCC around generating policies associated with reducing emissions from deforestation and forest degradation (REDD). These discussions originated under the climate policy regime as Avoided Deforestation but have since proliferated to influence the negotiations taking place under the Convention on Biological Diversity and discussions related to the role of the United Nations Forum on Forests in global policy processes, as the goal of REDD is to support maintaining forest vegetation that can operate as a carbon sink and to prevent the release of carbon from forest resources.

While engaged actors have developed various strategies for shaping policies and institutions associated with REDD, the conceptual currency associated with gender has provided an important tool for stakeholders who are concerned about the potential of REDD policies and institutions to ignore environmental and social safeguards. The demands of actors to ensure that safeguards be included in emerging REDD regimes have led to a shift in the discourse, reflected in the current standard phrase REDD+. The plus in REDD+ had originally become a stand-in for an elaboration of the need to ensure that social and environmental justice concerns be incorporated into regimes associated with the reduction of carbon emissions that are related to deforestation and forest degradation. More recently, as REDD+ became a more typical way of referring to the emerging REDD-related policy processes, and as stakeholders began to be concerned that safeguards were *still* being disregarded, stakeholders have made more explicit arguments for the inclusion of environmental and social justice mechanisms into REDD+ as well. This speaks to Harvey's concern that '[s]ophisticated discursive strategies are now in place... to absorb and defray the different imaginaries that typically root much of radical ecological thinking' (Harvey 1998: 174–175). Harvey speaks here of radical ecological thinking. Yet, critical analysis of the safeguards needed to ensure that REDD mechanisms do not merely prioritise carbon, thus exacerbating social inequalities, can be analysed in much the same way. While it is clear that civil society and Indigenous peoples' organisations (IPOs) have been pushing to include such safeguards in emerging REDD regimes, it is equally clear that this engagement with policy will continue to be an ongoing process, as the critiques that have been levelled at REDD policy regimes are perhaps addressed rhetorically rather than in practice. The recent publication of the International Law Principles for REDD+ by the Indian Law Resource Center (2012) is an indication of the international community's need to specify base-level standards as REDD+ programmes are being implemented. As Armstrong Wiggins writes in the introduction to the report, '[t]he REDD+ Principles were prepared in response to the need identified by indigenous peoples for international REDD+ actors to respect their human rights' (Crippa and Gordon 2012). Much of the politicisation of the REDD+ proposals and programmes has centred around the potential for negative environmental and social repercussions as REDD+ projects are implemented by various institutions and in the context of multiple nation states with varying extant legal frameworks in relation to land tenure and indigenous rights.

While gender and climate change are utilised in this chapter as a way of specifying the analysis and providing concrete examples, in particular related to the moves on the part of various civil society organisations (CSOs) to influence the unfolding REDD regimes, the notion of strategic use of conceptual currency to influence policy deliberations is applicable beyond this specific case. This analysis takes seriously the framework articulated by Cox associated with what he terms a critical theory that does not take institutions and social power relations for granted but calls them into question by concerning itself with their origins and how and whether they might be in the process of changing (1986: 208). In taking up questions related to the ways in which gender can be used strategically in shaping the transforming climate policy regime, a certain complexity develops. In many ways, the conceptual currency of gender is not entirely distinct from that of other stakeholder interests, as it has become ideologically entwined with notions such as participation and transparency that are also compelling and codified concepts within UN policy processes. There are clearly conceptual slippages that occur when the currency of gender is deployed in that it is, at times, taken to stand in for myriad social justice concerns that often remain unarticulated. However, this is precisely the nature of conceptual currency – the discursive and ideological assumptions that have become associated with particular terms such as gender are often not spelled out. Policy is replete with such ambiguity. Indeed, the unspoken and ideological components can lend legitimacy to particular strategic practices that members of CSOs deploy as they attempt to shape policy outcomes.

Negotiating gender: conceptual currency and strategic participation

Since the United Nations Conference on Environment and Development (UNCED) established specific intergovernmental bodies for negotiating policy related to climate, biological diversity and desertification in 1992, CSOs and IPOs have developed particular strategies for engaging with UN-based policy deliberations. The result of the 1992 Rio process was that CSOs and IPOs became more networked and organised around issues that were being framed in specific ways through the UNCED institutions. In addition, some CSOs, recognising the importance of the institutionalisation of the Rio agreements, reinvented themselves to directly engage in post-Rio negotiating processes. An example that epitomises this dynamic is the Women's Environment and Development Organisation (WEDO). Originally known as Women USA, WEDO shifted its mission in the early 1990s in large part in order to be more influential in international development and specifically to work jointly on a strategy for the UN Conference on Environment and Development (WEDO website). This is a significant shift as it also speaks to the transformation in both women's-based social movement organising and in academia to take into account a more global analysis of gender. As will be addressed later in this chapter, the transnationalisation of the women's movement came about, in part, due to a critique of second-wave feminism that was largely seen as articulating the interests of relatively privileged white women in the global North. Thus, the transformation from Women USA to WEDO in the early 1990s directly reflects this conceptual shift. It was a shift that was intended to address the critique that the category of women is not universal. Not all women experience marginalisation based on the same factors, and, indeed, women in the global North should not be setting the agenda for policy processes that deal with gender. However, the deployment of the term gender in policy negotiations that take place under the auspices of the United Nations often does not allow for a complex treatment of the disparities in privilege or marginalisation that women may experience – women who are differently situated geographically or in terms of structural dynamics associated with race, class, religion, sexuality or ethnicity, among other salient factors.

In spite of the broad use of the term gender without specification as to whom it references, based on concrete data regarding the results of gender inequalities in particular locations and based on significant social movement organising around problems associated with the marginalisation of women from resources – particularly in the global South – the notion of gender has been established as being a crucial part of UN negotiations processes. Indeed, the category of women constitutes one of the nine Major Groups codified in the UNCED document *Agenda 21*. The Commission on the Status of Women (CSW) has coordinated four global conferences related to women since 1975. Furthermore, it is clear that, in spite of specific policies and movements designed to address women's concerns, gender inequalities continue to persist on a variety of levels. In 2010, the UN General Assembly voted to establish the United Nations Entity for Gender Equality and the Empowerment of Women, which is intended to serve as an umbrella organisation within the institutional framework of the United Nations. This organisation has the explicit mandate to support policy processes related to women's concerns, including those that take place under the CSW.

Many of the women's organisations that became engaged with UN environmental policymaking in the 1990s had existing interests in international development. Others, like WEDO, shifted their explicit mission to address transnational and/or development concerns. In both sorts of cases, space had been carved out for women's organisations to legitimately engage in various policymaking processes – primarily under the rubric of inclusion of women's concerns in international development. Likewise, as climate concerns in general have become more dominant in UN policy processes, many women's development organisations have explicitly incorporated a treatment of climate issues into their missions. This is not to say that this is merely a strategic move to maintain relevance or access to resources. There is a wide body of academic and social movement-based literature that outlines the ways in which climate change is not a gender-neutral phenomenon (Dankleman 2010; Denton 2002; Glazebrook 2011). Thus, the increased focus on climate issues within policy fora also speaks to an increased urgency on the ground in specific locations as people attempt to cope with a changing environment. In order to facilitate networking among organisations as well as participation in policy processes, some organisations have been created that aim to enhance the institutional frameworks of existing women's organisations and to coordinate efforts related to climate change policy. For example, the organisation Climate Justice Now is intended as a network of organisations and movements from across the globe committed to the fight for social, ecological and gender justice (Climate Justice Now website). However, on the one hand, in the case of many organisations that address gender among other climate-related concerns, gender remains somewhat tangential or isolated from the primary focus of the organisational work. On the other hand, many organisations have sprouted from the traditional women's movement organisation to view climate and climate change through a gender lens with the particular goal of keeping both climate and gender in focus. This is the case for the Mary Robinson Foundation for Climate Justice (http://www.mrfcj.org).

A major locus of women's organising around the emerging climate regime can be found in the organisation GenderCC – women for climate justice, which is the global network of women and gender activists and experts from all world regions working for gender and climate justice (GenderCC website). With elaborate research tools and policy-related information, including an explicit segment entitled 'Mainstreaming Gender into International Climate Change Policy', GenderCC is expressly designed to provide a platform for a critical application of gender issues to climate policy debates and thus intervene in the unfolding climate regime. As a result of organising at UNFCCC meetings in the early 2000s, individuals involved with GenderCC, WEDO and other organisations with a gender and development focus successfully argued for a Women and Gender NGO constituency to be officially recognised by the UNFCCC. Thus, as CSOs who engage in UNFCCC meetings are delineated into topic areas (Research and Independent NGOs, or RINGOs, Environment NGOs, or ENGOs and so on), the Women and Gender NGO constituency was provisionally recognised by the UNFCCC in 2009. Otherwise known as the Women's Caucus, the institutionalisation of Women's and Gender NGOs in the UNFCCC system provides a more strategic point of entry for organisations interested in ensuring that the gender dynamics associated with climate change are recognised. In line with the analysis above regarding the discursive terrain of gender within the context of UN policymaking, the Charter of the Women and Gender Constituency explicitly states that its goal is to formalise the voice of the women's and gender CSOs present and regularly active in UNFCCC processes and to debate, streamline and strengthen the positions that these organisations put forth (Women and Gender Constituency Charter). Furthermore, it is explicitly stated that '[t]he Constituency draws upon global commitments to gender equality and women's rights, especially as they relate to climate change, and toward the achievement of the Millennium Development Goals (MDG) and related commitments and Conventions. The Constituency works to ensure human rights and a gender perspective is incorporated into UNFCCC negotiations, plans and actions' (ibid). The language of the Charter relies upon a historical establishment of the importance of gender as an element in policymaking processes. Additionally, it situates itself in relation to previously agreed-upon commitments, such as the MDGs. As such, it pulls in a range of known-in-common discursive schemes to legitimate the need for a specific integration of gender-focused organisations into the climate policy regime.

One of the very practical ways that GenderCC has engaged with the policy regime is to create a very readable and brief yet detailed pamphlet called Gender into Climate Policy: Toolkit for Climate Experts and Decision Makers (GenderCC 2009). The introduction to this pamphlet clearly presents the problem statement that '[g]ender is a critical and overlooked dimension in climate policy negotiations' (ibid). Furthermore, the authors go on to argue that '[w]hen the UNFCCC was agreed upon gender was not addressed.... However, integrating gender into policy making is now widely regarded as best practice in formulating climate policies and programs' (ibid). A key point is articulated here. The integration of gender into policymaking has become a significant element for achieving governance systems that are effective and participatory. For organisations that were involved in the 2010 UNFCCC meetings in Cancun, the incorporation of official language associated with gender into the texts was considered to be a significant success. The texts negotiated at Cancun contain eight references to women and gender across seven sections. Thus, the notion of gender contains substantial conceptual currency within UN policymaking, as it is associated with best practices for policy generation and also project-level implementation. However, the crux of the issue is that effective inclusion of gender means very different things to different people. As a key element of the conceptual currency of the UN, gender carries with it a range of social justice and equity-related assumptions and meanings. But the very nature of conceptual currency is that an ideological shortcut stands in for an elaboration of what is explicitly intended, thus allowing for conceptual or practical slippage.

Not only does the ambiguity in the language of policy present hazards in terms of the potential for a variety of interpretations of meaning, but, interestingly, the use of the term women or even gender without specification as to *which* women in *which* circumstances has the potential to gloss over important differences between women who are differently situated. Universalising categories such as women have been called into question by recent feminist scholarship and social movement organising. The transformation of the organisation Women USA to WEDO is a classic example of the results of this sort of critique. The assumptions and goals of relatively privileged second-wave feminism failed to address the multitude of experiences of women from a variety of racial, national, class and other backgrounds. The argument is that women in the global North should not set the agenda for climate policy as it will not reflect the multitude of experiences of women in the global South. Indeed, both social movement activist and academic arguments about the differential impact of climate change on women tend to focus on the vulnerability of women in the global South. Thus, it is recognised that the term gender may serve to gloss over important distinctions between women who are differently situated in terms of access to resources. Instead, '[d]ifferentiating factors such as class and caste, kinship, age, nationality, and socio-cultural group are important variables. Analysing these differences is as crucial as looking into the differences between women and men' (Dankleman 2002:24). However, in deploying the currency of the term gender in climate negotiations, much of this complexity is lost. Indeed, the strategic use of the term gender in pushing policy to be more inclusive often relies on the assumption that it references the concerns of women in the global South, particularly in terms of their vulnerability to issues associated with climate change.

Highlighting these conceptual pitfalls is intended to speak more to the process level of policymaking than to suggest that gender should not be strategically used in making policy. Clearly, it is imperative to continue to attempt to push policy processes to be as equitable and inclusive as possible. In the case of gender and climate change, changing local environments as a result of global climate changes will only continue to exacerbate gender disparities and marginalisation of women. However, policy often does not speak of these day-to-day realities as women are forced to travel further for water or work harder to produce subsistence foods, among myriad other repercussions of changing local environments. Indeed, the very nature of policy is that it is constructed around vague and ambiguous terms that are held together primarily by known-in-common interpretive schemes. This is essentially the crux of debates that have been unfolding regarding what it means for policy processes to be participatory or transparent or accountable (Hale 2008). Clearly, the scholarship on legitimacy (referenced by Cadman, Introduction) works towards defining and specifying some of those terms.

In some ways, the rhetoric of gender ensnares actors in particular identity politics that has the potential to perpetuate specific categorisations and abstractions (Bannerji 1995). As mentioned above, in many cases, the term gender is used in negotiations without clear specification as to what (or who) it intends. However, it is important to emphasise that this can also be utilised to strategic advantage by stakeholders in policy negotiations, as the process of policy deliberation allows for participants to utilise the ambiguity of terms and concepts in order to influence policy outcomes. The agreed-upon importance of gender equality and the conceptual linkages between gender and human rights allow stakeholders to pressure policymakers to incorporate more inclusive language in final texts. The ambiguity of the terms can provide openings for stakeholders to engage in policymaking regimes and to push those regimes to incorporate social justice dynamics.

As policy regimes are being formulated to address climate change, the incorporation of safeguards into policy processes is crucial to a range of stakeholders. Agostino and Lizarde note in their article on gender and climate justice that the concept of climate justice emerged as a result of introducing a rights approach to the challenges posed by climate change (2012: 90). They furthermore argue that the rights approach is not only imperative in creating equitable and effective climate-related policies, but it can also help to work towards a mechanism to bring about transformative justice (Agostino and Lizarde 2012: 95). The discourse of social movements associated with gender and climate change is increasingly incorporating a justice framework that further helps to incorporate a range of issues associated with marginalisation and power.

Conclusions and recommendations

While potentially problematic due to the lack of specificity that the term gender contains, the use of the conceptual currency that is associated with gender within UN-based policy processes can be a useful way of pushing policymakers to incorporate social and environmental safeguards into climate mitigation and adaptation regimes. Although there are potential issues inherent in the ambiguity of the term gender as it gets taken up in climate negotiations, the institutionalisation of gender into the United Nations as an agreed-upon concept allows for the potential to provide strategic leverage to a range of stakeholders who are concerned about the possibility for climate-related policies to exacerbate existing inequalities. However, it is also clear that stronger policies – in terms of both mitigation and adaptation – are being instituted on regional and local levels. Therefore, as Gale has pointed out in his chapter, civil society actors must continue to engage in multiple spheres of strategy work beyond the formal climate regime in order to ensure that the international policies are as robust as possible and also to ensure that what takes place on the ground does not exacerbate existing inequalities or further marginalise vulnerable people from essential resources.

References

Agostino, A. and Lizarde, R. 2012. Gender and climate justice. *Development*, 55 (1), 90–95.

Bannerji, H. 1995. *Thinking through: essays on feminism, marxism, and anti-racism.* Toronto, ON: Women's Press.

Cox, R., 1986. Social forces, states and world orders: beyond international relations theory. *In:* R. Keohane, ed. *Neorealism and its critics*. New York: Columbia University Press.

Crippa, L.A. and Gordon, G. 2012. *International law principles for REDD+*. Washington, DC: Indian Law Resource Center.
Dankleman, I. 2010, *Gender and climate change: an introduction*. London: Earthscan Press.
———. 2002. Climate change: learning from gender analysis and women's experiences of organising for sustainable development. *Gender and development*, 10 (2), 21–29.
Denton, F. 2002, Climate change vulnerability, impacts and adaptation: why does gender matter? *Gender and development*, 10 (2), 10–20.
Eastwood, L. E. 2005. *The social organization of policy: an institutional ethnography of UN forest deliberations*. New York: Routledge Press.
Escobar, A. 1995. *Encountering development: the making and unmaking of the third world*. Princeton, NJ: Princeton University Press.
Gender, C.C. 2009. *Women for Climate Justice Platform*, http://www.gendercc.net/policy/conferences/cop11.html, accessed 15 July 2012.
Glazebrook, T. 2011. Women and climate change: a case-study from northeast Ghana. *Hypatia*, 26 (4), 762–782.
Hale, T.N. 2008. Transparency, accountability, and global governance. *Global governance*, 14, 73–94.
Harvey, D. 1996. *Justice, nature, and the geography of difference*. London: Blackwell Publishers.
Lambrou, Y. and Pianna, G. 2005. *Gender: the missing component in the response to climate change*. Food and Agriculture Organization of the United Nations.
Rohr, U. 2009. *Gender into climate policy: toolkit for climate experts and decision-makers*. Berlin: GenderCC-Women for Climate Justice.
Smith, D.E. 1987. *The everyday world as problematic: a feminist sociology*. Boston, MA: Northeastern University Press.
Visvanathan, S. 1991. Mrs. Brundtland,s disenchanted cosmos. *Alternatives*, 16, 377–384.

4
Governing Adaptation Policies and Programmes

Geoff Cockfield

Introduction

Increasing greenhouse gas (GHG) emissions and slow progress on international emissions mitigation negotiations have encouraged the development of governance systems for adaptation to the impacts of climate change. Adaptation involves initiatives that moderate harm and/or exploit beneficial opportunities (McDonald 2010: 8). For this discussion, *autonomous adaptation* includes responses undertaken by individuals, families and businesses without incentives, requirements or exhortations from an authority, while *planned adaptation* involves some level of planning for a group, region, nation or group of nations (McDonald 2010: 8). Planned adaptation requires a range of policy instruments and actions, and responses can be reactive or anticipatory (examples in Table 4.1) and involve incremental or major change. For all but the least developed countries (LDCs), planned adaptation is more about government, especially national government, than governance at this stage, for reasons that will be explained in this chapter. The nature of the problem (climate change) and the diversity of possible adaptation responses mean however that more comprehensive governance arrangements, including some international coordination, will be both desirable and necessary.

Given the relatively recent interest in planned adaptation, compared to mitigation, it is premature to fully evaluate climate change adaptation governance, but it is possible to suggest how the quality of governance, as it develops, might be enhanced. This chapter starts with an argument that progress on and prospects for emissions mitigation are such that adaptation policy will likely become increasingly important. Then there is an overview of the development of the international and national

Table 4.1 Examples of adaptation issues, responses and policy domains

Biophysical effect	Impacts	Autonomous adaptation	Planned adaptation	Policy domains
Increasing number of high-temperature days/fewer low-temperature days	Heat stress	Relocation Housing insulation Changed work patterns	Relocation payments[I] Insulation regulations[R] Insulation subsidies[I] Air-conditioning subsidy[IM]	Health Building standards Regional development
	Changes in disease and pest vectors	Relocation Local pest/disease control	Pest/disease control regulations[R] Biotechnological research[E]	Health Biological sciences
	Changes in crop & pasture ranges and yields	New production methods Different crops/varieties Relocation of farmers Cease farming	Research & development[E] Production subsidies[I] Farm adjustment programmes[I]	Agriculture International trade Welfare
	Changes in habitat ranges	Species movement	Relocation of species Vegetation protection legislation[R] Payments for habitat protection[I]	Environment Natural resources management
Increased frequency and severity of floods	Threats to human life Property loss	Relocation of people Local flood barriers	Rescue and recovery services New insurance products Relocation payments[I] Land use regulations[R] Planned flood barriers	Disaster management Financial services Natural resources management

	Crop & livestock loss	Change production patterns	Research & development[E] Early warning systems	Agriculture
Reduced rainfall and run-off	Reduced irrigation water	New irrigation methods Different crops/varieties	Research & development[E] Regulate water capture and/or allocations[R] Create water markets[D]	Water management Agriculture
	Reduced domestic water	Changed consumption patterns	Consumption regulations[R] Water saving campaign Desalination/recycling plants	Water management Public engineering works
Higher sea levels	Property loss	Relocation of people	Rescue & recovery services Compensation payments[I] Land use regulations[R] Sea walls	Disaster management Land use planning Public engineering works

Notes: [R]Regulations; [I]Incentives; [D]Removal of disincentive (subsidised water); [E]Education and research; [M]Maladaptation.

structures for governing adaptation responses. This leads to the main argument that some degrees of national and international coordination will be both desirable and necessary but that participation, especially at the local and sectoral levels, will also be necessary given the high levels of uncertainty about both the diversity and diffusion of impacts as well as the consequent need for multi-level governance (Steurer et al. nd: 3). This discussion primarily focuses on developed countries where evaluation studies have been conducted but any discussion of international arrangements requires consideration of adaptation planning in some developing countries. This breadth of consideration means that national variations are not considered in detail, some of which can be found in more specific and comparative studies (see, for example, Biesbroek et al. 2010; Harmeling et al. 2011; GEF 2009; MFAD and Preston et al. 2011; Steurer et al. nd; Tompkins et al. 2010).

The limits of mitigation policies

In this section, it is argued that pressures for adaptation, both autonomous and planned, will increase because there will be insufficient outcomes from mitigation measures to prevent significant climate change. The reasons for this insufficiency include the dynamics of international climate change negotiations, here considered through the prism of game theory, the politics of climate change as an economic and social issue and concern about the economic impacts of mitigation. There is also the nature of the policy problem. Garnaut argues that climate change is a 'diabolical' problem, meaning that it is at the extreme of wicked problems (2008: xviii). It certainly fits some of the criteria of a wicked problem, as originally defined, in that there are many stakeholders, with divergent views and so there is unlikely to be a definitive judgement of the best solutions; it is a unique problem; experimentation is difficult because every attempt to solve the problem has its own consequences; and there are major consequences from choosing responses that do not work or work well enough (Rittel and Webber 1973: 162–167). According to Garnaut, it is 'uncertain in its form and extent ... insidious rather than (as yet) directly confrontational' and 'long term rather than immediate, in both its impacts and remedies' (Garnaut 2008: xviii). It therefore seems to fit with Head's (2008) notion of intensifying 'wickedness', where there is high systemic complexity; divergence of viewpoints and values; and uncertainty in relation to 'risks, consequences of actions, and changing patterns' (Head 2008: 103).

There is uncertainty about the rate of emissions, how emissions will affect the climate, the severity of socio-economic impacts and the effectiveness of mitigation measures (Mace 2006: 62; McDonald 2010: 27). There is uncertainty about the effect of GHG releases from vegetation and oceans (Anderson et al. 2008; Mann 2009), and offsetting aerosol effects from industrial activity and emissions data may be incomplete and rapidly outdated (Anderson et al. 2008: 3718; Mann 2009). Such uncertainties have been used, as an example, by President Bush (cited in Kolstad 2007: 69) to argue against signing up to the Kyoto Protocol (KP), but uncertainty is only one aspect of a complex policy environment. Since the atmosphere is a commons for all species and individuals and it is a public good, climate change is a 'collective-risk social dilemma' (Milinski et al. 2008: 2291). Many actions will be needed to reduce emissions and impacts and yet the benefits are uncertain and well in the future (Milinski et al. 2008: 2291). Furthermore, there are incentives for countries to free ride by not joining a coalition to address the problem and by not complying when in a coalition (Finus 2008; Kolstad 2007).

There are also differing incentive levels to participation because of the differences in costs and benefits between countries (Kolstad 2007: 69), and it is especially difficult to develop effective coalitions if major powers stand out of agreements (Huettner et al. 2010). De Zeeuw argues that 'in the most important situations, in which environmental damage costs are relatively substantial, large coalitions will not easily arise', and if they do, they will be 'unstable' due to defections (de Zeeuw 2006: 164–173). Carbone et al. (2009) argue that for progress to be made on international agreements on climate change mitigation, China needs to be in a coalition, preferably with the United States. In relation to the KP, there are major powers that are non-signatories and it is not self-enforcing on the signatories (Barrett 2006: 23). A more optimistic view is that the Conference of Parties (COP) is a framework that started as a loose agreement, with the prospect of firming up over time with members pushing others to join and then reciprocity leading to cooperation (Brekke and Johansson-Stenman 2008: 288–289; Huettner et al. 2010). Finus argues that even 'non-cooperative' arrangements, whereby participants join over time and targets are developed, are possible and may even be superior in terms of outcomes to more static cooperative agreements (2008: 33).

There is however little evidence from the COPs of the development of binding agreements with little progress at either Copenhagen (COP 15) or Cancun (COP 16) (Anderson and Bows 2011: 20–22; Garnaut 2011: 51). In a collective risk game, participants need to be convinced of either or both: that there is a high probability of dire consequences and that there is some 'fairness' about the contributions (Milinski et al. 2008: 2294). The major players are not yet experiencing or perhaps acknowledging dire consequences and there are very different views on the fairness of proposals. Developed countries are looking forward and wanting higher targets for developing countries while developing countries look back at the disproportionate contribution of the developed countries to current GHG accumulations (Huettner et al. 2010). This is not to argue that lack of progress on mitigation is inevitable since there might be game changing developments. For example, the evidence for 'dire consequences' may become more compelling and widely accepted or China might make a rapid policy shift, given its efforts on alternative energy (Garnaut 2011: 13, 53–55). The international negotiation 'game' is, however, just one factor that national governments consider.

Potential participants in agreements must also be mindful of national political factors, and mitigation policy is highly politicised in some – perhaps in all – countries. Pearse argues that there are coordinated and interlocking 'denial and delay' campaigns against action on mitigation in developed countries (2007: 213). Factors that contribute to resistance to action include the structure of news media, business interests and pre-existing ideological contests. In relation to media, there is the limited scientific knowledge of many journalists (Antilla 2005: 340) and a commitment to 'balance', which is enacted as the need to present an alternative, even if these views are very much those of a minority (Carvalho 2007: 233). There is also the desire for conflict (Carvalho 2007: 224). Antilla (2005: 350) found that US newspapers ran articles accepting the orthodox climate science while also running 'controversialist' arguments. Business interests also want to avoid or delay the financial costs of mitigation strategies and use lobbyists and public relations to those ends, and some have even funded alternative journals as an outlet for the contrarian 'science' (Antilla 2005: 340; Pearce 2010: 214). Also intersecting with business and the media are the ideological or cultural warriors, often from think tanks supported by business, who see climate change activism and even state action as a Green/Left attack on capitalism (Carvalho 2007: 230–231; Pearse 2010: 217). There are also economic 'pragmatists' who argue that there is likely to be warming but that major mitigation efforts would significantly retard economic growth and anyway technological solutions will arise and so there is no need for 'precipitous' and costly action (Carvalho 2007).

Positions among the general population also range from concerned to sceptical about the underlying theory and the costs of action

(Lorenzonia et al. 2007: 450). Emissions trends may concern scientists and environmentalists but there are other trends that influence popular perceptions. As an example, one regular survey in the United States reveals that the proportion of respondents who were 'concerned' or 'alarmed' about climate change decreased in the wake of the global financial crisis (Leiserowitz et al. 2011: 12). Public opinion can also turn quickly with particular weather events, such as major rainfall (Schneider and Lane 2006: 46). Then there is the nature of climate change as a perceived risk. According to Marks et al., risks are more acceptable if they are 'visible, voluntary, familiar, controllable, fair, forgettable, and that the adverse effects will be immediately experienced rather than delayed' (2008: 85). In the case of climate change the phenomenon is invisible, there is involuntary exposure to risks, the impacts of which are unevenly distributed, and many effects seem to be well in the future. Some surveys suggest that a significant proportion of people do not feel they have enough information to have an opinion (Lorenzonia et al. 2007: 450). This suggests some scope for change with the accumulation of more evidence, but even with widespread understanding this does not mean there is concern and even concern does not necessarily lead to behavioural change (Lorenzonia et al. 2007). Other perceptions that impede individual responses include the scale of the problem, the temporal or geographic distance of impacts, perceptions that it is too late to bother with individual action, not wanting to be one of few taking action and a reluctance to make personal sacrifices (Lorenzonia et al. 2007: 450).

With the limited progress on international negotiations and the political and economic imperatives for nations to pursue growth, there are signs that carbon emissions are tracking at the higher end of projections or even exceeding them (Anderson et al. 2008: 3715). Parry et al. (2009) propose that peak emissions in 2015 and a 3 per cent annual reduction thereafter would result in a temperature peak of 2°C in 2065 but that a ten-year delay in peak emissions would result in an average increase of 2.5°C. Yet, some growth scenarios suggest up to a six-fold (with high economic growth) increase in emissions to 2100 (Blanford et al. 2009: 583). In the European Union (EU), there was little progress on emissions reduction up to 2005, yet it comprises countries that should be leading the way (Saikku et al. 2008). Population growth is relatively low, environmental regulation is widely accepted, vegetation mass is increasing and there is an emissions trading scheme, though compulsion is limited (Saikku et al. 2008). If progress on mitigation is slow where the inclination to regulate capitalism is greater, then this does not bode well for more libertarian regimes or developing countries with a focus on improving quality of life. According to Anderson and Bows there is 'now little to no chance of maintaining the rise in global mean surface temperature at below 2°C... [M]oreover, the impacts associated with 2°C have been revised upwards' (2011: 40). Even a 2°C increase will result in major changes to ecosystems and food production, and as more data come in and analytical techniques develop, the risk levels from even 2°C warming are being upgraded (Anderson et al. 2008; Smith et al. 2009: 4137). Hence at the very least, autonomous adaptations are inevitable, since the lack of stringent mitigation measures will necessitate action.

The emergence of adaptation governance

Planned adaptation with national and some international coordination can be justified on the grounds of the threat of maladaptations and the values of efficiency and equity. Relying solely on autonomous adaptation would increase inequities between developed and developing countries, between regions within nations, current and future generations and humans and other species less able to adapt (Schneider and Lane 2006: 28–29). Climate change impacts will also likely disproportionately affect the vulnerable, especially the poor and less mobile within societies (Adger et al. 2006: 6). Then there is the problem of autonomous adaptations that are maladaptations in that they contribute to the initial problem, such as an increased use of air-conditioning resulting in increased energy use and emissions. Planned adaptation may also be more efficient than cumulative autonomous adaptations since state agencies can put in place regulations or incentives for action when impacts are less obvious but costs are lower (Tompkins et al. 2010). Biesbroek et al. (2010: 442) found that some EU countries were developing adaptation structures and processes because of the recognition of the costs of inaction but other drivers included extreme weather events, NGO advocacy and business and media concern, and developments in the United Nations Framework Convention on Climate Change (UNFCCC).

The UNFCCC adaptation architecture developed in three stages. The initial focus of that organisation from the 1990s to 2000 was on mitigation, although Article 4 of the Framework Convention required Parties to cooperate to prepare for adaptation and develop plans for vulnerable regions (Ayers et al. 2010: 273; United Nations 1992: 5–9). In the second 'phase', 2001–2007, there was increasing recognition of the need for adaptation, especially in the 'developing' and 'least developed' countries

(LDCs) (Ayers et al. 2010: 273). From the 2001 Marrakesh Conference of Parties (COP 7), there was special attention to countries with vulnerable areas and particular difficulties under mitigation policies (UNFCCC Conference of Parties 2002: 32). Specific programmes included the Least Developed Countries Fund (LDCF) to support National Adaptation Programmes for Action (NAPA) and the Kyoto Protocol Adaptation Fund (KPAF) (Ayers et al. 2010: 273–274). There is also a Strategic Priority on Adaptation under the Global Environment Facility (GEF) to strengthen research and development, integrate 'climate change considerations into sustainable development planning' and enhance capacity, including 'institutional capacity' (UNFCCC 2002: 34–35).

The concerns about trends in emissions and climate change, as discussed above, were acknowledged in the IPCC's 2007 report and at the Bali Conference (COP 13) it was resolved to enhance 'action on adaptation' to build resilience, manage risks and develop disaster reduction strategies (UNFCCC 2008: 4). In Cancun in 2010 (COP 16), Parties agreed to form an Adaptation Committee, and in Durban in 2011 (COP 17), the composition of the Committee was agreed. The Parties also agreed that National Adaptation Plans (NAPs) should 'enable all developing and developed country Parties to assess their vulnerabilities, to mainstream climate change risks and to address adaptation' (UNFCCC 2012: 80). Developed countries were urged to financially support the LDCs through the LDCF and developing countries (excluding the listed LDCs) were 'invited' to use the NAP guidelines to formulate their own plans (UNFCCC 2012: 82–83).

During the second stage a number of developed countries produced adaptation plans for sectors, cities and regions (see for examples from the United Kingdom, the United States and Australia Preston et al. 2011: 429–430; Tosehe Natural Resource Management Ministerial Council 2004; The Natural Resource Management Ministerial Council 2006), and Finland and Spain developed national plans (Biesbroek et al. 2010: 441). From 2007, the number of sectoral and local/regional plans increased considerably with Australia, the United Kingdom and at least another six EU countries producing national plans (Biesbroek et al. 2010; Council of Australian Governments 2007; Preston et al. 2011). In the United States, there is an Interagency Climate Change Adaptation Taskforce to provide policy recommendations (Preston and Westaway 2011: 408). Adaptation planning documents vary in origin, scope and objectives. For example, from a comparison of Australia, the United Kingdom and the United States, some plans are from one government agency, others cut across multiple agencies and just more than half prescribe particular adaptation actions (Preston and Westaway 2011: 415). Biesbroek et al. (2010: 443) compared nine national strategies of EU countries, including the United Kingdom, and found some commonalities, but there are a wide range of approaches to the planning. Much of the early focus is on adaptation research, rather than actual adaptations (Biesbroek et al. 2010; Council of Australian Governments 2007), but there are some evaluations of plans and institutions (Biesbroek et al. 2010; MFAD and GEF 2009; Preston et al. 2011; Tompkins et al. 2010) and these provide insights into how outcomes might be improved.

Adaptation may be appealing to national policy actors frustrated with, or wary of, international negotiations, uncertain about the degree and impacts of climate change and concerned about domestic politics and economics, but planned adaptation is not necessarily an easy option if a medium- to longer-term perspective is taken. Indeed, climate change with an increasing focus on planned adaptation would in some ways make for greater policy wickedness than climate change with mitigation as the main focus. There are the same uncertainties in relation to emissions trends and the consequent impact on the climate, but there is increased complexity because of the diversity of effects, the difficulty in downscaling climate impact models and the interaction of climate change with other problems (McDonald 2010: 24–25). Most impacts, such as those from Table 4.1, interact with and exacerbate other problems (McDonald 2010: 11). For example, there is a range of threatening processes that affect habitats, including agriculture, urbanisation and pest species, so changes in temperature and rainfall are additional stressors. Given these interactions, the range of impacts to be addressed and the uncertainty of timing, extent and severity of impacts, there is a strong case for planning for adaptation governance, rather than relying on ad hoc responses.

The elements of adaptation governance

Multi-level and multi-spatial governance of climate change mitigation will become more solidly institutionalised because of the nature of the problem, and the institutional arrangements for adaptation will necessarily be different than those for mitigation. As outlined and summarised in the Figure 4.1, national governments in developed countries are at the centre of the emerging governance of adaptation. As described in the previous section, the UNFCCC and other international agencies are guiding planning in the LDCs through particular programmes and funding (solid arrows in Figure 4.1), but so far it is only recommended

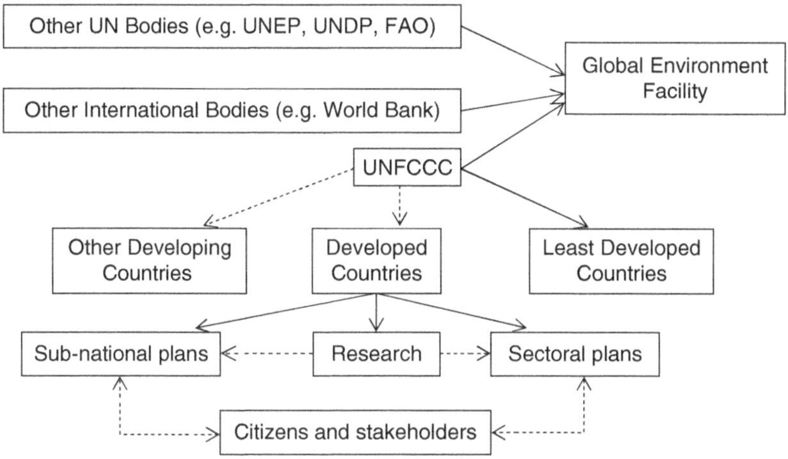

Figure 4.1 Emerging governance of adaptation structures and programmes
Note: Black arrows indicate some degree of formal governance; dashed lines indicate recommendations, aspirations and/or voluntary interactions.

that developing (excluding the defined LDCs) and developed countries work to produce NAPs. As will be argued below, there may well be more developed international adaptation governance architecture (increasingly solid lines) as the non-LDC countries address cross-border issues. In the meantime, national governments will be at the centre of adaptation because of their institutional position and economic and political power.

Most obviously, national governments are central because they have the funding and political authority by way of elections and habits of obedience. Second, they are primary actors in shaping the national political economy and the alignment of policy instruments with the particular political economy. For example, in Australia where market liberalism has a strong influence, one might expect the development of market-based adaptation instruments. These could include new insurance products, enhancing markets for water, ensuring free trade in food and paying for conservation initiatives likely to contribute to adaptation (Garnaut 2011: 102–108). In a different jurisdiction, say a country in the northern EU, regulatory instruments might be more of an option. The third reason for national governments being at the centre of adaptation planning is that they usually control or at least fund the organisations with the capacity to contribute to the adaptation research, such as national scientific organisations. Fourth, they can most comprehensively determine how matters of equity are considered and this will involve balancing efficiency, state finances and vulnerability, as opposed to just exposure, to climate change impacts (Adger et al. 2006: 13; Macintosh 2010: 45). Fifth, planned adaptation will require horizontal coordination because the effects will cut across conventional policy domains (as in Table 4.1) (Weber and Khademian 2008). Finally, national governments will be at the centre of vertical coordination, from the international down to the local level. Major mitigation policies, such as internationally coordinated emissions trading schemes, theoretically need only two levels of governance, with internationally agreed targets and national implementation. For adaptation, geographically small countries such as Singapore could perhaps also manage with a two-level adaptation framework but for larger countries the uncertainty, diversity and distribution of effects will require local issue identification and implementation (Biermann and Boas 2010: 224; Macintosh 2010: 61). There would also need to be local involvement of stakeholders, for programme success will depend on the acceptance by the affected group of the definition of the issues, that change is required and that the options are acceptable (McDonald 2010: 12).

National governments will be involved in international adaptation negotiations and arrangements and there are several justifications for this. Biermann and Boas (2010: 226) argue that some issues, such as climate change refugees (see Chapter 13, this volume), food availability and human security, require international coordination. Also, the UNFCCC coordinates the transfer of funds from developed countries to the least developed countries through the adaptation programmes described above. The international forums provide an opportunity for sharing adaptation lessons that some argue will be critical to the required adaptive management at national and local levels. For some countries there will be trans-boundary issues, such as water use planning, requiring negotiations, that may be facilitated by a recognised international body (Harmeling et al. 2011). As described above, adaptation programmes and discussions are emerging from the 'institutionalized arena of transnational global climate governance' (Stripple and Pattberg 2010: 142). These developments are rather tentative, since the statements from the COP on adaptation are shot through with language, 'urging' and 'advising' members to take action, while developed countries are concerned about commitment given the uncertainty about impacts, the possibilities of maladaptation and wanting to make adaptation part of mainstream development (Mace 2006: 62). Ayers et al. (2010: 270) argue that adaptation has also been seen as the 'poor cousin' of mitigation and as a competing option. Some countries have been slow to pay contributions and there have been difficulties in identifying the climate change parts of the various social and economic problems (Ayers et al. 2010: 276–277). Coordination across a range of international bodies, such as the FAO and the World Bank, is also relatively weak at this stage (Biermann and Boas 2010: 227–231). Membership of the Adaptation Committee, though covering different types of countries, is still very state orientated and nominees are meant to have expertise in adaptation. Ayers et al. (2010: 277) argue that this strong focus on expertise and technical solutions can mitigate against 'bottom-up approaches rooted in community-based patterns of resource management that contribute to building resilience'.

Biesbroek et al. also found this to be a problem at the national level in their review of EU adaptation plans (2010: 448). They argue that there are likely to be greater challenges in multi-level governance and policy integration than in finding technical solutions. They advocate for a 'strong, leading department, ministry or institution' to develop and manage a national strategy (Biesbroek et al. 2010: 446). An example of one such coordinating agency is the Australian Department of Climate Change and Energy Efficiency (DCCEE), but such agencies can be vulnerable to bureaucratic change. The DCCEE was formerly the Australian Greenhouse Office and has been moved between departments and threatened with various structural changes by the opposition parties. Biesbroek et al. (2010: 446) also advocate for '(sub)units on adaptation in leading vulnerable sector departments', presumably referring to departments such as agriculture and water resources. Interdepartmental units could address some of the horizontal coordination issues – as would the use of agreed adaptation 'assessment instruments', such as environmental impact assessments (Biesbroek et al. 2010: 447). Further to that the UK Climate Impacts Programme has developed specific adaptation analysis tools (Tompkins et al. 2010: 629).

Preston and Westaway (2011: 409) argue that in general there will need to be coordination of the lessons learnt from adaptation policies and projects, given the need for adaptive management. Without some coordination and social learning, adaptive management can become ad hoc incrementalism as they found in their comparative study (Preston and Westaway 2011: 427). They see benefits in the formalised NAPA process that provides templates for national plans, though they also note the problem of underfunding of these. They also argue that there was more focus on 'knowledge acquisition and capacity building measures' than on specific actions (Preston and Westaway 2011: 428). Specific adaptation projects, while risky due to uncertainty of outcomes and difficulties in measuring programme impact, at least start to provide data for adaptive management. Specific projects can also be ranked according to timing, with those with long lead times being given priority (Macintosh 2010: 61). Tompkins et al. conclude from their study of the United Kingdom that actual adaptations are 'rarely undertaken solely in response to climate change' and one important driver is apparent co-benefits (2010: 633). The potential for co-benefits can help gain the support of stakeholders concerned with other issues, especially if governments take a 'no-regrets' approach (Macintosh 2010: 61; McDonald 2010: 12). There is a special category of co-benefits in which there are both mitigation and adaptation benefits. Adaptation and mitigation are substitutes at the global scale but not necessarily at the local scale (Macintosh 2010: 39). For example, increased use of building insulation as an adaptation to higher temperatures may reduce air conditioning requirements and the consequent emissions.

An ideal future landscape of adaptation governance would see a UNFCCC Adaptation Committee that works with other international agencies to coordinate funding and programmes for developing countries, serves as a repository for best practices in adaptation practices and governance, undertakes research on adaptation technology and practices relevant to the international issues described above and involves environmental and development NGOs to identify issues and incorporate advice. National governments would then provide the adaptation research framework, a central coordinating agency, adaptation sub-units in functional areas and forums for working with sub-national governments or arrangements. For an example of the latter, Australia has for long used the Council of Australian Governments to develop uniform approaches to particular issues across a federal system and adaptation could be another of those issues. There will however be considerable variation across countries given differences in geographic and economic scales, the power of sub-national governments and the political culture of state intervention. By the nature of the problem and solutions and the history of the emergence of adaptation governance, national governments will remain the key determinants of the policy framework, but outcomes will be inhibited unless broader and effective governance arrangements develop.

Conclusions and recommendations

The international architecture for planned adaptation is emerging, from COP talks, NAPs and programmes for LDCs. There are potential problems with this architecture in that the slow progress on emissions control and inter-country conflicts may also constrain progress on adaptation. Indeed, already the language around adaptation is reminiscent of the mitigation 'game' with each COP negotiation resulting in more exhortations for countries to develop plans, but nothing is binding, except for some financial leverage over the LDCs. There is therefore a case for separating adaptation from mitigation negotiations, but on balance it is better to try and work with the architecture that is emerging.

There is a strong case for having a central adaptation agency, but it needs to have some guaranteed institutional stability and authority to work across other portfolios. Such an agency would be strengthened by having a budget for adaptation sciences even if working through existing national research agencies and some control over sectoral funding, dependent on adaptation reporting and performance in other portfolios. This agency would also ideally be the repository for information on, and evaluation of, local and sectoral adaptation experiments, and given the uncertainties, the thinking in terms of experimentation will be extremely important. In federations or confederations (the EU), adaptation planning will also be strengthened if there is ongoing planning around adaptation through formal meetings, supported by continuing work at the relevant agency level.

Then there are the issues of who should be in the governance processes and how they should be consulted. With mitigation there are some obvious organisations that can be brought into consultative and deliberative processes, such as environmental groups, peak industry groups and scientific and economic experts to formulate and implement national policies, such as emissions trading schemes and carbon taxes. Adaptation requires many actions across many areas and domains. Permanently established NGOs and local organisations may not have the capacity to deal with such diversity and specific-purpose solutions and could struggle to participate in deliberations over time. Indeed, the nature of adaptation planning, being experimental and ongoing, will see highly changeable policy communities. For example, there might be an early focus on land use planning to address rising sea levels that will see developers and local governments and communities intensely involved in regional plans for a time. Then there might be similar efforts with building regulations, water use and so on that will involve different groups. If, as argued here, one of the best strategies for getting action is in emphasising co-benefits, then this will effectively involve recruiting different groups into the particular policy community and avoid the temptation to repackage existing unrelated foreign aid as adaptation assistance.

References

Adger, W., Paavola, J. and Huq, S. 2006. Towards Justice in Adaptation to Climate Change. *In:* W. Adger, J. Paavola, S. Huq and M. Mace, eds. *Fairness in adaptation to climate change*. Cambridge, MA: The MIT Press, 1–19.

Anderson, K. and Bows, A. 2011. Beyond 'dangerous' climate change: emission scenarios for a new world. *Philosophical transactions of the royal society a*, 369 (1934), 20–44.

Anderson, K., Bows, A. and Mander, S. 2008. From long-term targets to cumulative emission pathways: reframing UK climate policy. *Energy policy*, 36 (10), 3714–3722.

Antilla, L. 2005. Climate of scepticism: US newspaper coverage of the science of climate change. *Global environmental change*, 15 (4), 338–352.

Ayers, J., Alam, M. and Huq, S., 2010. Global adaptation governance beyond 2012: developing-country perspectives. *In:* F. Biermann, P. Pattberg and F. Zelli, eds. *Global climate governance beyond 2012: architecture, agency and adaptation*. Cambridge: Cambridge University Press, 270–285.

Barrett, S. 2006. Climate treaties and 'breakthrough' technologies. *American Economic Review*, 96 (2), 22–25.

Biermann, F. and Boas, I. 2010. Global adaptation governance: setting the stage. *In:* F. Biermann, P. Pattberg and F. Zelli, eds. *Global climate governance beyond 2012: architecture, agency and adaptation*. Cambridge: Cambridge University Press, 223–234.

Biesbroek, G., Swart, R., Carter, T., Cowan, C., Henrichs, T., Mela, H., Morecroft, M. and Rey, D. 2010. Europe adapts to climate change: comparing national adaptation strategies. *Global environmental change*, 20 (3), 440–450.

Blanford, G., Richels, R. and Rutherford, T. 2009. Feasible climate targets: The roles of economic growth, coalition development and expectations. *Energy economics*, 3 (Supplement 2), 582–593.

Brekke, K. and Johansson-Stenman, O. 2008. The behavioural economics of climate change. *Oxford review of economic policy*, 24(2), 280–297.

Carbone, J., Helm, C. and Rutherford, T. 2009. The case for international emission trade in the absence of cooperative climate policy. *Journal of environmental economics and management*, 58 (3), 266–280.

Carvalho, A. 2007. Ideological cultures and media discourses on scientific knowledge: re-reading news on climate change. *Public understanding of science*, 16 (2), 223–243.

Council of Australian Governments. 2007. *National climate change adaptation framework*. Canberra: Department of Climate Change and Energy Efficiency, http://www.climatechange.gov.au/government/initiatives/national-climate-change-adaptation-framework.aspx, accessed 16 January 2012.

de Zeeuw, A. 2006. Dynamic effects on the stability of international environmental agreements. *Journal of environmental economics and management*, 55 (2), 163–174.

Finus, M. 2008. Game theoretic research on the design of international environmental agreements: insights, critical remarks, and future challenges. *International Review of environmental and resource economics*, 2 (1), 29–67.

Garnaut, R. 2011. *The Garnaut review 2011: Australia in the global response to climate change*. Melbourne: Cambridge University Press.

Garnaut, R. 2008. *Climate change review: final report.* Melbourne: Cambridge University Press.

Harmeling, S., Kreft, S. and Rai, S. 2011. *Institutions for adaptation: towards and effective multi-level interplay*. Bonn: Germanwatch and WWF International.

Head, B. 2008. Wicked problems in public policy. *Public policy*, 3 (2), 101–118.

Huettner, M., Freibauer, A., Haug, C. and Cantner, U. 2010. Regaining momentum for international climate policy beyond Copenhagen. *Carbon balance and management*, 5 (1), http://www.cbmjournal.com/content/5/1/2, accessed 16 January 2012.

Kolstad, C. 2007. Systematic uncertainty in self-enforcing international environmental agreements. *Journal of environmental economics and managament*, 53 (1), 68–79.

Leiserowitz, A., Maibach, E., Roser-Renouf, C. and Smith N. 2011. *Global warming's six Americas, May 2011*. Yale Project on Climate Change Communication, New Haven, CT: Yale University and George Mason University.

Lorenzonia, I., Nicholson-Cole, S. and Whitmarsh, L. 2007. Barriers perceived to engaging with climate change among the UK public and their policy implications. *Global environmental change*, 17 (3–4), 445–459.

Mace, M. 2006. Adaptation under the UN Framework Convention on Climate Change: The International Legal Framework. *In:* W.N. Adger, J. Paavola, S. Huq and M. Mace, eds. *Fairness in adaptation to climate change*. Cambridge, MA: The MIT Press, 53–76.

Macintosh, A. 2010. A theoretical framework for adaptation policy. *In:* T. Bonyhady, A. Macintosh and J. McDonald, eds. *Adaptation to climate change: law and policy*. Sydney: The Federation Press, 38–62.

Mann, M. 2009. Defining dangerous anthropogenic interference. *Proceedings of the national academy of science*, 106 (11), 4065–4066.

Marks, J., Martin, B. and Zadoroznyj, M. 2008. How Australians order acceptance of recycled water: National baseline data. *Journal of sociology*, 44 (1), 83–99.

McDonald, J. 2010. Mapping the legal landscape of climate change adaptation. *In:* T. Bonyhady, A. Macintosh and J. McDonald eds. *Adaptation to climate change: law and policy*. Sydney: The Federation Press, 1–37.

MFAD (Ministry of Foreign Affairs of Denmark) and GEF (Global Environment Facility). 2009. *Operation of the least developed countries fund for adaptation to climate change*. Copenhagen: Ministry of Foreign Affairs of Denmark.

Milinski, M., Sommerfeld, R., Krambeck, H-J., Reed, F. and Marotzke, J. 2008. The collective-risk social dilemma and the prevention of simulated dangerous climate change. *Proceedings of the national academy of sciences*, 105 (7), 2291–2294.

Parry, M., Lowe, J. and Hanson, C. 2009. Overshoot, adapt and recover. *Nature*, 458 (7242), 1102–1103.

Pearce, F. 2010. *The climate files: the battle for the truth about global warming*. UK: Random House.

Pearse, G. 2007. *High & Dry*. Melbourne: Penguin Group (Australia).

Preston, B., Westaway R. and Yuen, E. 2011. Climate adaptation planning in practice: an evaluation of adaptation plans from three developed nations. *Mitigation and adaptation strategies for global change*, 16 (4), 407–438.

Rittel, H. and Webber, M. 1973. Dilemmas in a General Theory of Planning. *Policy sciences*, 4 (2), 155–169.

Saikku, L., Rautiainen A. and Kauppi, P. 2008. The sustainability challenge of meeting carbon dioxide targets in Europe by 2020. *Energy policy*, 36 (2), 730–742.

Schneider, S. and Lane, J. 2006. Dangers and Thresholds in Climate Change and the Implications for Justice. *In:* W.N. Adger, J. Paavola, S. Huq and M. Mace, eds. *Fairness in adaptation to climate change*. Cambridge, MA: The MIT Press, 23–51.

Smith, J., Schneider, S., Oppenheimer, M., Yohe, G., Hare, W., Mastrandrea, M., Patwardhan, A., Burton, I., Corfee-Morlot, J., Magadza, C., Fussel, H-M., Pittock, A., Rahman, A., Suarez A. and van Ypersele J-P. 2009. Assessing dangerous climate change through an update of the Intergovernmental Panel on Climate Change (IPCC) 'reasons for concern'. *Proceedings of the national academy of science*, 106 (11), 4133–4137.

Steurer, R., Bauer A. and Feichtinger J. nd. *Analytical framework for the stock taking of governance arrangements in 10 OECD countries*. Vienna: Institute of Forest, Environmental, and Natural Resource Policy.

Stripple, J. and Pattberg P. 2010. Agency in global climate governance: setting the stage. *In:* F. Biermann, P. Pattberg and F. Zelli, eds. *Global climate governance beyond 2012: architecture, agency and adaptation*. Cambridge: Cambridge University Press, 137–145.

The Natural Resource Management Ministerial Council. 2004. *National biodiversity and climate change action plan 2004–2007*. Canberra: Department of Sustainability, Environment, Water, Population and Communities.

The Natural Resource Management Ministerial Council. 2006. *National agriculture & climate change action plan 2006–2009*. Canberra: Department of Agriculture, Fisheries and Forestry.

Tompkins, E.L., Adger, W.N., Boyd, E., Nicholson-Cole, S., Weatherhead, K. and Arnell, N. 2010. Observed adaptation to climate change: UK evidence of transition to a well-adapting society. *Global environmental change*, 20 (4), 627–635.

United Nations 1992. *United Nations Framework Convention on Climate Change*. No location: United Nations.

UNFCCC (United Nations Framework Convention on Climate Change) Conference of Parties. 2012. *Report of the conference of the parties addendum part two: action taken by the conference of the parties in its seventeenth session*. Durban: United Nations.

UNFCCC (United Nations Framework Convention on Climate Change) Conference of Parties. 2002. *Report of the conference of the parties on its seventh session*. Marrakesh: United Nations.

UNFCCC (United Nations Framework Convention on Climate Change) Conference of Parties. 2008. *Report of the Conference of the Parties Addendum Part Two: Action taken by the Conference of the Parties in its thirteenth session*. Bali: United Nations.

Weber, E. and Khademian, A. 2008. Wicked problems, knowledge challenges, and collaborative capacity builders in network settings. *Public administration review*, 68 (2), 334–349.

5
Applying an Empirical Evaluation to the Governance Legitimacy of Carbon Offset Mechanisms on the Basis of Stakeholder Perceptions

Timothy Cadman

Introduction

This chapter begins with a background to the various carbon offset mechanisms, public and private, within the climate change regime complex. It continues with a quantitative and qualitative analysis of stakeholder perceptions regarding the governance of the UN climate change negotiations on reducing emissions from deforestation and forest degradation in developing countries (REDD+). Governance was evaluated by means of a series of online surveys conducted during the period 2009–2011, using the framework of principles, criteria and indicators developed by Cadman (2011) and presented in the Introduction to this volume. Respondents were selected from state (that is, governmental) and non-state (that is, non-governmental) interests and further separated by their geo-political location in either the 'global North' or 'global South'. The results show that survey respondents generally found REDD+ to be inclusive but did not consider that there was the necessary capacity or resources for meaningful participation. A concluding section reviews the framework applied and comments on the nature of multi-stakeholder relations in contemporary global governance and REDD+ specifically.

Background: carbon offset mechanisms

Many of the climate change mechanisms can be described as a market-based attempt to 'decouple' environmental degradation from economic growth by advancing the notion that industrial capitalism can be made more 'environmentally friendly' (Bäckstrand and Lövbrand 2006: 53). They can be characterised as existing along a continuum of market models: from technocratic and neo-liberal solutions such as clean development with no alteration of existing institutional and governmental structures to those with a greater emphasis on ecological democracy, institutional reform and reduced state centrism (ibid). One of the most commonly known and publicly recognised market-based instruments are carbon credits. These relate to emission-based allowances as well as project-based offsets. Carbon offsets are a subset of credits that allow targets for emissions reduction to be achieved in one place through the purchase of emission reductions from a project, located somewhere else. The regulated or compliance market was established by the Kyoto Protocol (KP) and is implemented via the Clean Development Mechanism (CDM). The voluntary offset market runs in parallel and is a mix of non-governmental and private sector-driven schemes. Both are regulated differently. They are however interlinked since they evolved from a common base in the early 1990s and because they are often retailed together, with 'failed' or delayed compliance offsets being sold on the voluntary market. The CDM was established under the 1997 KP, part of the United Nations Framework Convention on Climate Change (UNFCCC), which includes a number of carbon offset governance mechanisms. The CDM has an Executive Board and related CDM-Methodology and Accreditation Panels as well as small-scale working groups. Corporations and non-governmental organisations (NGOs) also have governance roles. The CDM establishes quality standards through complex, lengthy and staged project methodologies relating to project design, method approval, validation, registration and verification, after which the project is issued with a Certified Emissions Reduction (CER) offset (Lovell 2010: 353–354).

In terms of governance, the CDM could be described as a 'second generation' example of environmental problem solving, typified by its voluntary, market-based and 'partnership' approach. Although it provides for increased non-state influence, including NGOs, this collaborative philosophy is not seen as being particularly evident in policy practice, due largely to government–business networks that are more concerned with flexibility than poverty reduction. Here the carbon offset market is to be seen as reinforcing the neo-liberal governance vision shared by business and government in developing and industrialised countries. State centrism and technocratic attitudes are also reinforced in the verification requirements, which favour managerial and scientific policy elites (Bäckstrand and Lövbrand 2006: 69–71).

The CDM has been portrayed as a 'North/South bargain to facilitate technology transfer, include developing countries in emissions trading mechanisms, and crucially, harness the abilities and resources of the private sector to implement innovative, cost-effective emissions reductions projects through a global market mechanism' (Bumpus and Cole 2010: 541). Despite this, its real contribution to sustainable development has been questioned, even if it is undeniable that it transfers money to pay for emissions reductions (ibid). The necessity of some level of reform to take the mechanism away from 'project-based' carbon reductions towards 'grand' technology transfer initiatives that can lower transaction costs and create wider benefits has been recognised. COP 15 requested the development of top–down, broadly applicable and standardised baselines to be applied to CDM projects, aimed at reducing emissions via Nationally Appropriate Mitigation Action (NAMA). The UN Programme to Reduce Emissions from Deforestation and Forest Degradation (REDD/REDD+) is an example of this move. National financing mechanisms will however need regular reporting, and countries will need to accept greater levels of accountability, possibly through external verification and review. In the case of REDD+, for example, robust and accurate baselines of deforestation are essential (Macey 2009: 447–448).

Voluntary offsets were developed independently of Kyoto by NGOs, businesses, charities and individuals and constitute a 'parallel' market, with a greater variety in governance arrangements and practices than the compliance market. There is no equivalent of the CDM Executive Board, and the market consists of competing standards with different assessment criteria, ranging from the need for local participation to reduction measurement and verification. These offsets are often referred to as Voluntary Carbon Offsets (VCOs) or verified emissions reductions (VERs), but there are other terms in the market as well. VCOs *contra* CERs are seen as having a greater focus on sustainable development issues – including poverty alleviation – in the global South (Lovell 2010: 354–357). VCO projects were first developed in the 1990s and were few in number, but are now in the hundreds. By 2008 transactions equivalent to 123 metric tonnes of carbon took place. This growth can be attributed to the 'pre-compliance' offsets – delayed or rejected products in the official compliance market – entering the voluntary system. They cater mostly to the non-Kyoto signatory countries and regions of the United States (28 per cent), Middle East (45 per cent) and Asia (45 per cent); this is to be compared with the CDM where 84 per cent of the CER market is in China. There were 17 VCO standards in 2008. The most well known is the Voluntary Carbon Standard with 48 per cent of the market share and a price of US$7.34 per credit in 2008, the Gold Standard, developed by environmental NGO WWF, which has 12 per cent of the market share and a price of $14.40 in 2008. It is focused exclusively on energy efficiency and renewable energy and is perhaps the most stringent in terms of rules for local stakeholder consultation (Lovell 2010: 354–359).

VCOs have been criticised for a lack of quality in evaluation, favouring 'tick-box' outputs; even the Gold Standard with its emphasis on stakeholder consultation is lacking in this regard. Essentially, the standards deliver outcomes but not *processes* of negotiation, and they do not properly account for or really tackle sustainable development in their implementation at the local level (Bumpus and Cole 2010: 543). In 2009, the UK government established a non-mandatory Quality Assurance Scheme (QAS), developed with stakeholders, for voluntary offset providers. Lovell views this action as 'an example of how a public authority...has attempted to reassert its power in the largely NGO and privately governed offset market' (Lovell 2010: 360). However, most key providers have decided not to apply for accreditation, rejecting the standard and rendering it irrelevant (Lovell 2010: 360).

REDD+

At the 11th Conference of Parties (COP) in Montreal in December 2005, the UNFCCC first introduced into negotiations an agenda item on reducing emissions from deforestation in developing countries. At the Bali climate negotiations in 2007, the initial concept was expanded, acknowledging that 'forest degradation also leads to emissions, and needs to be addressed when reducing emissions from deforestation' (UNFCCC 2008). The subsequent '+' was formally added as a consequence of discussions at COP 15 (7–18 December 2009), with the aim of placing greater emphasis on protecting and enhancing the carbon values of forests as well as reducing emissions (Parker et al. 2009). REDD+ is best understood as an attempt to use the market to reduce greenhouse gas (GHG) emissions associated with forest clearing, by allowing 'avoided deforestation' to be included in carbon trading mechanisms. It is effectively a payment in exchange for actively preserving existing forests (Barbier and Tesfaw 2012). Negotiations via the UNFCCC COPs are ongoing and the mechanism is not yet finalised, although a number of demonstration projects and national-level 'readiness' activities are underway (Cerbu et al. 2011). REDD+ is linked to the KP and its related CDM, offering developed countries a way to meet their emissions targets by reducing GHG emissions and increasing GHG sinks (that is, forests) in developing countries (CIFOR nd).

Effective implementation of REDD+ could contribute to a 50 per cent reduction in global deforestation by 2020 and zero per cent by 2030 (Angelsen et al. 2010). By 2030, forest-related mitigation could contribute to more than a third of all reductions in CO_2 emissions. Should this occur, it will be essential to have strong national arrangements for verification and certification, including the checking of the accuracy of the data reported (IUCN 2010). Developed countries provided US$4 billion to REDD+ for the funding period 2010–2012 (Forest Carbon Portal 2010). In May 2010, Norway pledged US$1 billion to assist Indonesia to reduce its GHG emissions from deforestation and forest degradation. Representatives of Indigenous peoples and local communities will take part in the planning, implementation and institutional management of the funds allocated for REDD+ in Indonesia (Royal Norwegian Embassy in Jakarta 2010). However, given the lack of clarity over land tenure and the lack of explicit recognition of Indigenous peoples' rights in the agreement, doubts have been raised over the credibility of these claims (Lang 2010). Such issues as Indigenous peoples' rights – and ensuring the rights of interested and affected parties generally – confront REDD+ with a number of governance challenges. As an example of multi-level, multi-actor governance, from the intergovernmental to the local and including a wide range of interests, it has been argued that if REDD+ is to address local concerns and global problems it requires an institutional architecture that is both participatory and deliberative. Where there is a wide range of actors functioning within cross-sectoral partnerships (of which there are now many at the national REDD+ project level), addressing issues of inclusiveness, equality and capacity building as well as accountability and transparency are critical to effective collaboration. Decision-making also requires new political processes for reaching agreement. These should embrace both formal and informal negotiating procedures, and address disputes equitably, should they arise (Forsyth 2009: 113–122).

Method

The analysis below explores stakeholder perceptions regarding the extent to which REDD+ is seen to be addressing these governance challenges. Four surveys were deployed using Survey Monkey (www.surveymonkey.com) to assess the perceptions of state and non-state actors, from both the global North (developed countries) and global South (developing countries). Participants were recruited from publicly available lists of email contacts of organisational representatives active in global environmental negotiations relating to REDD+. Survey 1 was a preliminary feasibility trial of approximately 500 recipients. The database of recipients was further refined for Survey 2, when further stakeholders were identified and included, and others were removed from the list on request, totalling approximately 600. Surveys 3 and 4 were sent to this list. An Internet link to a pre-designed questionnaire was emailed to recipients, but owing to anonymity provisions it was not possible to determine exactly who responded to each survey. Respondents identified as either 'government' (that is, state), or 'environmental' and 'social' (that is, non-state, and specifically, non-governmental). Government respondents included representatives from relevant ministries, such as environment and forestry, and with responsibility for climate change policy. Environmental and social interests represented a range of civil society groups and comprised international and national environmental non-governmental organisations (ENGOs) and organisations with a more social orientation, including Indigenous peoples' organisations (IPOs).

Perceptions of the governance of REDD+ were evaluated using the 11 indicators discussed in the Introduction. Respondents were asked to rate their perceptions anonymously, using a Likert scale: 'very low' (0–1); 'low' (1.1–2); 'medium' (2.1–3); 'high' (3.1–4); and 'very high' (4.1–5). A Likert scale is widely used in assessing the perception of respondents through structured questionnaire surveys, in which respondents specify their level of perception on a given statement (in this case for an indicator). The average of all respondents in a given category captures the collective perception for the indicator within that subset of survey respondents. A time series of repeated surveys gives a series of averages that signifies the trends in perceptions for the indicators, by which the performance of each indicator over time can be determined (Burns and Burns 2008). This method of assessment is considered a balanced approach to analysis of perceptions because there are equal amounts of positive and negative attributions on the scale. Ordered response choices from seven to nine are occasionally used, although five ordered response levels, as selected for this study, are common (Dawes 2008). These often reflect choice-text using such terms ranging on a scale, from 'strongly agree' to 'strongly disagree', that is, from the positive to the negative. The approach adopted here for the survey is not typical, as the choice-language is designed to address perceptions of governance *quality* (from very high to very low) rather than agreement/disagreement. 'Low' and 'very low' are not strictly negative positions, but in the context of governance quality these can be interpreted as negative positions. But the overall average value of 'medium' shows a neutral perception on the indicators.

The results from the surveys were analysed using Statistical Package for the Social Sciences (SPSS). Opportunities for qualitative comment were also provided for each indicator, along with the option for further general comment at the end of the survey. The ratings at the indicator scale were aggregated under the relevant criterion; in turn the relevant criteria were combined to provide a result at the principle level. These principle-level results were added to provide an overall 'legitimacy score', out of 55. A conventional pass/fail target value of 50 per cent has been applied to the overall score. A target value represents a reference point to aim for (Lammerts van Bueren and Blom 1997: 24). The comprehensive results across all 11 indicators and associated criteria and principles are produced in Table 5.1.

Survey 1 was conducted during November 2009 (that is, before COP 15). Six governments (one from the North and five from the South) and twenty-six NGOs (twenty-one from the North and five from the South) responded. Survey 2 took place after COP 15, in March 2010. Thirteen governments (three from the North and ten from the South) and twenty-four NGOs (six from the North and eighteen from the South) responded. Survey 3 was conducted during September 2010 (that is, before COP 16 in Cancun, 29 November–10 December). Sixteen governments (seven from the North and nine from the South) and thirty-two NGOs (eleven from the North and twenty-one from the South) responded. Survey 4 was opened for a more extensive period from mid-November 2011 to mid-January 2012 to allow survey respondents the opportunity to respond before, during and after COP 17 (Durban, 28 November–9 December 2011). Eleven governments (three from the North and eight from the South) and forty-one NGOs (eight from the North and twenty-three from the South) responded.

Table 5.1 Breakdown of results of perceptions regarding UNFCCC REDD+-related negotiations among governmental and non-governmental stakeholders

Principle	1. Meaningful Participation (Maximum score: 25; Minimum: 5)								2. Productive deliberation (Maximum score: 30; Minimum: 6)									
Criterion	**1. Interest representation (Maximum score: 15 Minimum: 3)**				**2. Organisational responsibility (Maximum score: 10 Minimum: 2)**				**3. Decision-making (Maximum score: 15 Minimum: 3)**				**4. Implementation (Maximum score: 15 Minimum: 3)**					
Indicator	**Inclusiveness**	**Equality**	**Resources**	**Criterion Score**	**Accountability**	**Transparency**	**Criterion Score**	**Principle Score**	**Democracy**	**Agreement**	**Dispute settlement**	**Criterion Score**	**Behavioural change**	**Problem solving**	**Durability**	**Criterion Score**	**Principle Score**	**Total (out of 55)**
								November 2009										
State North (1)	5	4	1	*10*	3	4	*7*	17	4	4	4	*12*	4	4	5	*13*	25	**42**
State South (3)	3.2	2.6	2.2	*8*	3.3	3.2	*6.5*	14.5	3.2	3	2.5	*8.7*	2.6	2.4	2.8	*7.8*	16.5	**31**
Non-state North (21)	2.6	1.8	1.5	*5.9*	2.8	2.8	*5.6*	11.5	2.3	2.3	2.2	*6.8*	2.9	2.5	3.4	*8.8*	15.6	**27.1**
Non-state South (5)	2.6	3	1.8	*7.4*	2	2.4	*4.4*	11.8	2.3	2	2	*6.3*	2.6	3	3	*8.6*	14.9	**26.7**
								March 2010										
State North (3)	3.3	4	1	*8.3*	3	3.7	*6.7*	15	3.3	2.7	2.3	*8.3*	3.7	3.7	3	*10.3*	18.7	**33.7**
State South (10)	3.5	3.1	2.3	*8.9*	3.3	2.8	*6.1*	15	2.8	3.4	2.8	*9*	3.7	3.5	3.6	*10.8*	19.8	**34.8**
Non-state North (6)	2.7	1.8	1	*5.5*	2.6	2.7	*5.3*	10.8	2.2	2.2	2.2	*6.6*	2.6	2.6	2.7	*7.9*	14.4	**25.2**
Non-state South (18)	3.3	2.8	2.2	*8.2*	3.5	3.4	*6.9*	15.1	2.8	2.9	2.6	*8.3*	3.1	2.9	3.4	*9.4*	17.8	**32.9**

									September 2010									
State North (7)	4.1	3.7	2.5	10.3	3.4	3.4	*6.9*	17.2	4.1	3.2	3.2	*10.5*	3.9	3.6	3.6	*11.1*	21.6	**38.8**
State South (9)	3.9	3.7	1.8	9.4	3.1	3.5	*6.6*	16	3.5	3.2	3.3	*10*	4	3.9	4.3	*12.2*	22.2	**38.2**
Non-state North (11)	3.4	2.6	1.6	7.5	2.5	2.4	*4.9*	12.4	2.6	2	2.4	*7.2*	2.8	3	3.1	*8.9*	16.1	**28.5**
Non-state South (21)	4.2	3.6	2.2	10	3.7	3.9	*7.6*	17.6	3.7	3	2.7	*9.5*	3.8	3.7	3.5	*11*	20.4	**38**
									November–January 2011–2012									
State North (3)	3.5	3.7	2	*10.3*	3	3.7	*6.7*	17	3.7	4	3.7	*11.3*	4.3	4.3	4	*12.7*	24	**41**
State South (8)	4	3.5	2.6	*10.1*	3.5	3.9	*7.4*	17.5	3.8	3.8	3.5	*11*	3.8	3.9	3.9	*11.5*	22.5	**40**
Non-state North (8)	3.1	3.3	3	*9.4*	3	3	*6*	15.4	3	2.9	2.6	*8.5*	3	3.3	3.1	*9.4*	17.9	**33.3**
Non-state South (23)	3.5	3.2	2.1	*8.9*	3	3	*6*	14.8	3	2.7	2.7	*8.4*	3.3	3.2	3	*9.4*	17.8	**32.7**

Results

Quantitative analysis

Overall, the results appear to demonstrate that stakeholder perspectives regarding the governance quality and legitimacy of the UNFCCC REDD+-related negotiations have been relatively positive to date, with an average overall score of 31 out of 55. There were both similarities and differences in respondents' perceptions of REDD+. In terms of the final scores, Northern non-state respondents were the most negative, 'failing' REDD+ on two occasions (Surveys 1 and 2), awarding one narrow 'pass' (Survey 3) and providing only one definitively 'good' result (Survey 4). Southern non-state respondents rated REDD+ slightly higher, with three clear passes (Surveys 2, 3 and 4), but giving a 'fail' score in Survey 1. Southern state respondents scored REDD+ higher again, providing a 'pass' in all four surveys and in one instance (Survey 3) marginally exceeding the score of Northern state respondents. It is worth noting that Southern state and non-state respondents and Northern governments scored REDD+ very similarly in Survey 3. Of these three groups, Northern governments provided the highest 'pass' scores on three occasions (Surveys 1, 3 and 4). There is a clear trend in results: Northern and Southern non-state respondents usually provided the lowest scores, but Southern non-state respondents also tracked closely with Southern governments. However, Northern governments were the smallest group of respondents, and while all the results should be treated cautiously, it is especially true in their case.

At the indicator level, the trend for 'low' to 'high' (or 'very high') ratings generally followed a similar pattern among respondent groups. However, it is worth noting that there was an obvious 'low' cluster for the resources indicator among all groups. Previous research undertaken suggests this is a feature in other areas of global environmental governance and forestry in particular that can impact on the efficacy of both inclusiveness and equality (Cadman 2011: 181). Dispute settlement also performed relatively weakly. It has also been identified as a common problem in global environmental governance (ibid: 185–186). It is interesting to note, however, that inclusiveness generally received a 'high' or 'very high' rating among Northern state, Southern state and non-state respondents. Views among respondents regarding the equality of participation in UNFCCC REDD+-related negotiations is more mixed – perhaps reflecting the governmental orientation of negotiations at this level of REDD+. Overall, however, the general disaffection among the Northern non-state respondents in all four surveys should be noted.

Qualitative analysis

In terms of the ability to which Northern governmental interests could get their views represented, a comment was made that 'a lot depends on the negotiator and the chair'. For equality, one respondent felt that the various Parties were treated 'maybe too equally', adding that 'some countries have more interest at stake than others in REDD+'. For decision-making, one view was expressed that the effectiveness of REDD+ decision-making had become 'definitely less after Copenhagen'. The democratic arrangements underpinning UNFCCC decision-making led this respondent to opine that they were 'not so sure anymore that the one country, one vote system is adequate'.

One Southern governmental respondent was pleased by the inclusiveness of the interests of their specific country within the climate negotiations. They noted that it had been possible 'in cooperation with other (REDD) countries' to make sure that REDD+ was included in the Copenhagen Accord as well as 'several other multilateral programmes and in the development [of] financial mechanisms'. Views regarding equality were less positive, which was reflected in one respondent's comment that they felt that 'the intention is there', but they 'sometimes doubt[ed] the genuine willingness of some (developed) countries'. Concerning decision-making, the same respondent commented that because REDD+ was consensus-based – 'taking into account the complexity of this global issue and the different interests' – decision-making took a long time. They thought that this could be improved. However, the conclusion reached by this same respondent concerning democracy was that REDD+ was democratic only 'on paper'. They had 'noticed that some developed countries (with more resources and possibilities and political and economic power) [were] making the decisions, or influenc[ing] the decision making'. This same individual also felt that this was reflected in the manner in which disputes were settled.

One Northern ENGO who rated REDD+ 'high' for inclusiveness commented that 'up to Copenhagen this rating applie[d]'; 'post Copenhagen' they believed 'there [would] be lots of attempts to marginalise ENGO input'. According to another Northern NGO respondent, their ability to influence negotiations was 'probably high in terms of visibility, but medium in terms of effectiveness, as interests must be adopted by government to be officially on the table'. According to another, the lack of inclusiveness made the 'process very difficult to follow, not to mention influence, when you are not a government representative'. In terms of the inclusion of specific needs, another respondent felt that matters of 'environmental and social governance' were not being 'adequately addressed' in REDD+. This same respondent saw that the failure to address issues of importance only to specific interests meant that these concerns were not being accorded equal treatment; they pointed to forest-based communities in this regard. One respondent was particularly emphatic about the lack of equality. REDD+ negotiations were 'not equal. Vested interests and power plays mean[t] that certain, far less logical interests remain, despite large opposition'. Democracy also did not perform at all well. According to one respondent, 'the negotiating space is not equitable, for many well documented reasons, and power, corporate lobbying and vested interests have a greater impact on decision making than democratic processes'. Another made the interesting observation that 'UN processes seem inherently ineffective but would be even less democratic if they were more effective, I suppose'. Concerns about the making of agreements were also raised, with one respondent questioning the value of the agreements reached if there were no 'strong monitoring and enforcement mechanisms'. This same respondent also reflected on the problems associated with decision-making in the UN system as a whole. They commented on the fact that the UN process of 'requiring consensus among countries that may not truly be interested in protecting forests or forest-dependent communities... ha[d] serious drawbacks'.

Southern NGO responses were more numerous – and diverse – in their perspectives. One NGO respondent did not believe civil society could influence REDD+ negotiating processes. They raised the fact that 'only parties c[ould] actually participate, and they look[ed] out for their national interests'. Nevertheless, there were 'various forums which spill[ed] into the UNFCCC-REDD+ process' even if the 'official "process" itself d[id] not allow much for this'. In terms of inclusiveness, one respondent commented that to the extent that issues important to NGOs were heard, the degree of inclusion was medium to high, but in terms of the extent to which they were heeded the answer was low. For another, the problem was that only a 'few countries le[d] the political process of the negotiations'. To be inclusive, one respondent added that it was important to include marginalised groups in society, such as women. In terms of equality, another respondent challenged the notion that non-governmental interests should be treated equally. For them 'the process [was] driven by the COP, and [was] set up to vet the views of Parties and to provide a forum for them to reach agreement on a post-2012 climate change regime'. Another felt that REDD+ tried to treat everyone equally, but it was 'very hard to please everyone'. Views regarding democracy were similarly mixed. A respondent from Latin America felt that 'large countries include[d] provisions that did not favour the environment, but economic interests'; by contrast another felt that the REDD+ was democratic, but nevertheless added that 'the process by consensus [was] very inefficient and time consuming'. Views regarding the making of agreements were generally positive, although one respondent made an interesting observation that 'enterprise interests of the "West" almost always triumph', although this was 'mitigated with "noise" from the developing world'.

Discussion

There is a high level of similarity regarding the strengths and weaknesses identified by survey respondents. The 'low' rating for resources is common, no doubt reflecting the reality that few stakeholders are provided the sufficient wherewithal to participate in global governance. There is very little capacity at present in the mechanisms to address this issue either, given that the UN *Rules of Procedure* themselves do not address this issue. This is clearly a shortcoming in global governance: if participation is to be meaningful, greater attention needs to be paid to supporting under-resourced stakeholders and increasing the capacity of those who can afford to attend. In terms of the specific indicators investigated above, it is interesting to note that inclusiveness rates relatively well. This may indeed reflect the participatory developments in governance at the global level that have occurred since Rio and over the last 20 years. It might even be fair to say that inclusiveness is now normatively embedded, even if there are still a number of other governance shortcomings. In the case of REDD+, these reflect the highly state-oriented nature of the mechanism, participatory rhetoric notwithstanding: equality is a weak performer (indicating who really has weight in negotiations), as is dispute settlement (again indicating the existence, perhaps, of certain veto coalitions). More remarkable, however, is the fact that survey respondents from the global South generally rated the mechanism higher than their counterparts in the North. This is consistent across governmental and non-governmental respondents. Secondly, there is a similar difference in the gap between governmental and non-governmental respondents in the sense that governments consistently rated the mechanisms higher. These two trends are generally also reflected in the specific indicators of inclusiveness and equality and in the criterion of decision-making. It might be assumed that environmental and governmental stakeholders would be opposed to each other's perspectives in global environmental negotiations: governments often deliver less than is often hoped for, while NGOs usually demand tougher action. Although there is a discrepancy between the scores accorded to the mechanism investigated by these sectors, there is a general correspondence in overall perceptions: where Northern governments rate REDD+ lower than their Southern counterparts, so too do Northern NGO respondents. Southern governments and NGOs also have a gap in their rating, but again, there are correspondingly higher results across both sectors. In this regard, it seems that the nature of the perceptions of governance quality of Northern and Southern governments and their NGO opposites is one of 'parallel convergence' of opinion rather than divergence. The difference is to be found in the degree of confidence in REDD+. It might be expected to see governments' views converge with governments and the same for NGOs, but it is possible that geo-political factors come into play, rather than sectoral affiliation. An alternative or complementary hypothesis might be that the Southern sectors are beneficiaries in all three mechanisms, while Northern governments, as donor countries, contribute to implementation. NGOs from the North are less likely to gain financially, and therefore have less of a stake in rating REDD+ highly.

Methodological caveats

A number of observations need to be made at this point regarding the results presented above. Respondents rated REDD+ at the indicator level. A question has arisen in similar studies of global governance as to whether qualitative data can be applied to develop quantitative results. The conversion of verbal descriptions (low, medium, high) into numerical results can lead to judgements hidden underneath the apparent authority of a number, and the pretension of exact science (Nanz and Steffek 2005: 373). In this study, there has also been an aggregation of scores at the indicator level to produce results at both the criterion and principle levels and the overall score. In some instances, this has resulted in a situation where REDD+ performed poorly in one indicator but met the threshold at the criterion level, or did not meet the criterion-level threshold, while still meeting requirements at a principle level. These factors should be taken into account when looking at the 'pass/fail' results for each survey. It must also be stated that the numbers of respondents are low. This may have resulted in an outlier effect, where very low numbers of respondents in one group can overly influence the results (particularly, this may be the case in the responses of government – North). Online surveys have generally lower participation rates than other forms of survey technique (Van Selm and Jankowski 2006: 447). It should be added that in many developing countries, the Internet is unreliable and subject to frequent interruptions due to power shortages, which can discourage participation. Nevertheless, completion rates were excellent, at 100 per cent, with the exception of the first survey, where 56 commenced the survey and 32 completed (a retention rate of 57 per cent). Nevertheless, with low response rates and in the absence of detailed information on non-responses, it is not possible to determine accurately how representative the answers are for a particular category. Even if trends are consistent, this may be a consequence of auto-selection by respondents who chose to answer the survey in a particular way that is not indicative of the stakeholder groups as a whole. It is therefore not possible to make representative claims, but the results presented are useful for exploratory analysis.

Conclusions and recommendations

Consequently, taking all these factors into consideration, this study should therefore be seen as a work in progress requiring further research to determine whether the trends identified are correct across the 'universe' of stakeholders involved in the REDD+-related UNFCCC negotiations. Nevertheless, it has provided a relatively nuanced framework of evaluation based on a complex of institutional arrangements (such as interest representation and decision-making) for 'good', governance.

REDD+ comprises multi-actors and is an ideal institutional venue in which to examine stakeholder perceptions of global environmental governance. Given the mix of actors in REDD+, the methods of participation and deliberation associated with negotiations are as important as the decisions made about a given issue. In view of the collaborative and 'partnership'-based model of REDD+ on the ground, there are increased expectations about the role of non-state actors, and this has created some dynamic tensions in the relations between non-state and state actors in the formulation of REDD+ policy at the international level. Southern state and non-state interests will increasingly benefit from REDD+ projects in developing countries. But there is something of a 'divide' in perceptions between Northern non-state actors and other respondents. This may have implications for negotiations as they evolve, and more REDD+ projects are implemented on the ground.

However, care should be exercised in drawing any definitive conclusions on the basis of the results provided in these surveys. Greater levels of participation across sectors would yield more certainty regarding the perceptions of each sector and sub-sector concerning the legitimacy of REDD+ as an example of global environmental governance.

References

Angelsen, A., Brown, S., Loisel, C., Peskett, L., Streck, C., and Zarin, D. 2009. *Reducing emissions from deforestation and forest degradation (REDD): an options assessment report.* Washington, DC: Meridian Institute.

Bäckstrand, K. and Lövbrand, E. 2006. Planting trees to mitigate climate change: contested discourses of ecological modernization, green governmentality and civic environmentalism. *Global environmental politics,* 6 (1), 50–75.

Barbier, E. B. and Tesfaw, A. T. 2012. Can REDD+ save the forest? The role of payments and tenure. *Forests,* 3 (4), 881–895.

Bumpus, A. and Cole, J. 2010. How can the current CDM deliver sustainable development? *Wiley interdisciplinary reviews: climate change,* 1 (4), 541–517.

Burns, A. and Burns, R. 2008. *Basic marketing research.* Second edition. New Jersey: Pearson Education.

Cadman, T. 2011. *Quality and legitimacy of global governance: case lessons from forestry.* Basingstoke: Palgrave Macmillan.

Cerbu, G., Swallow, B. and Thompson, D. 2011. Locating REDD: a global survey and analysis of REDD readiness and demonstration activities. *Environmental science and policy,* 14 (2), 168–180.

Centre for International Tropical Forestry (CIFOR). nd. *Simply REDD: CIFOR's guide to forests, climate change and REDD.* Bogor: CIFOR. http://www.cifor.cgiar.org/publications/pdf_files/.../MediaGuide_REDD.pdf, accessed 1 October 2010.

Dawes, J. 2008. Do data characteristics change according to the number of scale points used? An experiment using 5-point, 7-point and 10-point scales. *International journal of market research,* 50 (1), 61–77.

Forest Carbon Portal. 2010. *Multilateral interim REDD+ partnership established in Oslo,* http://www.forestcarbonportal.com/resource/interim-redd-partnership-established-oslo, accessed 2 October 2010.

Forsyth, T. 2009. Multilevel, multiactor governance in REDD+: participation, integration and coordination. *In:* A. Angelsen, ed. *Realising REDD+: national strategy and policy options.* Bogor: CIFOR, 113–122.

International Union for the Conservation of Nature (IUCN). 2010. *REDD+: scope and options for the role of forests in climate change mitigation strategies.* http://cmsdata.iucn.org/downloads/redd_scope_english.pdf, accessed 21 May 2010.

Lovell, H. 2010. Climate policy and action 'underneath' Kyoto and Copenhagen: China and the United States. *Wiley interdisciplinary reviews: climate change,* 1 (3), 353–362.

Lammerts van Bueren, E. and Blom, E. 1997. *Hierarchical framework for the formulation of sustainable forest management standards.* Leiden: The Tropenbos Foundation.

Lang, C. 2010. Norway-Indonesia forest deal: US$1 billion dollars worth of continued deforestation? *REDD Monitor,* 28 May 2010, http://

www.redd-monitor.org/2010/05/28/norway-indonesia-forest-deal-us1-billion-dollars-worth-of-continued-deforestation/, accessed 7 October 2010.

Macey, A. 2009. Climate change: governance challenges for Copenhagen. *Global Governance*, 15, 443–449.

Nanz, P. and Steffek, J. 2005. Assessing the democratic quality of deliberation in international governance: criteria and research strategies. *Acta politica*, 40 (3), 368–383.

Parker, C., Mitchell, A., Trivedi, M. and Mardas, N. 2009. *The little REDD+ book*. Oxford: Global Canopy Programme.

Royal Norwegian Embassy in Jakarta. 2010. *Norway-Indonesia REDD+ partnership – frequently asked questions*, http://www.norway.or.id/Norway_in_Indonesia/Environment/-FAQ-Norway-Indonesia-REDD-Partnership-/, accessed 7 October 2010.

United Nations Framework Convention on Climate Change (UNFCCC). 2008. Report of the conference of the parties on its thirteenth session, held in Bali from 3 to 15 December 2007, unfccc.int/resource/docs/2007/cop13/eng/06a01.pdf, accessed 6 April 2012.

Van Selm, M. and Jankowski, W. 2006. Conducting online surveys. *Quality & quantity*, 40 (3), 435–456.

6
Evaluating the Clean Development Mechanism

Tek Narayan Maraseni

Introduction

Climate change, caused by anthropogenic greenhouse gas (GHG) emissions, could result in increasing temperatures, rising sea levels and consequently changes in land use, such as movement of the zone of grain production away from the equator at a rate of 10 km/yr (IPCC 2007; Quiggin and Horowitz 2003: 439). With increased temperatures, many respiratory and cardiovascular diseases, coral bleaching and frequencies of heat waves and cyclone and extreme precipitation could increase (Preston and Jones 2006: 26). Concerns about such impacts led to the United Nations Framework Convention on Climate Change (UNFCCC) in 1992 and the subsequent related Conferences of Parties (COPs).

The Kyoto Protocol (KP), agreed at COP 3 in 1997, was a milestone in the development of global climate policy, as for the first time the majority of developed countries accepted responsibility for past emissions and caps on GHG emissions. The KP set legally binding GHG reduction targets of at least 5.2 per cent of 1990 levels by the first commitment period (2008–2012). To achieve this collective target, each developed country (Annex B countries of the KP) was given a specific GHG emissions target. The European Union (EU) agreed to reduce emissions by 8 per cent, Canada and Japan by 6 per cent while Australia was allowed to increase GHG emissions by 8 per cent of 1990 levels in the first commitment period. To meet this target in a cost-effective manner, the Protocol adopted three 'flexible' market-based mechanisms: Emissions Trading, Joint Implementation (JI) and the Clean Development Mechanism (CDM).

Of the three, the CDM is the only mechanism that links developed and developing countries and was the first global environmental investment and credit scheme of its kind. But it has been criticised for failing to contribute to sustainability in target countries (Sterk and Wittneben 2006: 285; Teravainen 2009: 180). There is a belief that the CDM has also helped a number of countries to reduce emissions. However, this has largely been within the EU as a consequence of its own Emissions Trading Scheme (ETS) (World Bank 2009: 1). The impact outside this region is questionable. This chapter, following Cadman (2011), discusses the CDM's performance, focusing on two central elements of implementation: problem solving and behaviour change. The problem-solving capacity of the mechanism is investigated by looking at the degree to which the CDM actually contributes to its stated objective of sustainable development. It finds that this objective is overshadowed by the provision for 'unilateral' projects under the mechanism and the inclusion of measures to reduce hydrofluorocarbons (HFCs) and nitrous oxide. With regard to behaviour change, an examination of the actual degree to which GHG emissions have been reduced by implementation of projects in-country shows that change is so low as to be insignificant. The chapter begins by looking at the architecture and adoption of the CDM, continues with a review of its contribution to sustainable development and emissions reduction, and closes with some conclusions and recommendations.

History, architecture, scope, proliferation and critiques of the CDM

The CDM was established under the 1997 KP and includes a number of governing bodies, including the CDM Executive Board and related CDM-Methodology and Accreditation Panels, as well as small-scale working groups. The two-fold objective of the CDM is to assist developing countries (non-Annex I countries) to achieve sustainable development and allow the Annex B ('traditional industrialised') countries to count emissions reduction outcomes from projects in developing countries towards meeting Kyoto targets (Sterk and Wittneben 2006: 272). The consequence is that the CDM helps developed countries meet their quantified emissions reduction obligations at lower cost, while also helping developing countries to achieve sustainable development. The CDM provides an opportunity to developing countries to make progress on managing climate change and development and addressing local environmental issues rather than being preoccupied with more pressing economic and social needs. There are three types of CDM projects: unilateral, bilateral and multilateral. Initially, CDM projects were designed to be bilateral and multilateral. In a bilateral design, an Annex B entity (listed in Annex B of KP) directly invests in the project and in return receives Certified Emissions Reductions (CERs), while in a multilateral approach, Annex B entities receive CERs proportionally to their share in the investment (Baumert et al. 2000: 4). Finally, as of 2005, CDM projects can be developed and implemented by host countries unilaterally and the resulting CERs can be sold on the carbon market (Michaelowa 2007: 19). One CER is equivalent to one tonne of CO_2 emissions. As of 15 June 2011, 3,169 CDM projects had been registered in 71 developing countries and these projects were anticipated to generate more than 2,010 million CERs by the end of 2012 (UNFCCC 2011).

Since the registration of the first CDM project in late 2004, the CDM market has grown exponentially. The overall value for the CDM market was US$2.6 billion in 2005, US$6.2 billion in 2006, US$12.8 billion in 2007 and US$32.8 billion in 2008 (World Bank 2008: 1 and 2009: 1). The demand for CDM projects is largely driven by the CDM's eligibility to enter the European Union Emissions Trading Scheme (EU ETS), the world's largest carbon market) and on account of other voluntary and national carbon markets.

On the supply side, the first CDM project was registered by Brazil in 2004. Since then, project approvals have increased exponentially: 62 in 2005, 409 in 2006, 426 in 2007, 431 in 2008, 684 in 2009 and 735 in 2010 (UNFCCC 2011). Of the total registered projects, more than 36 per cent are from China, 23 per cent from India, 8 per cent from Brazil, 6 per cent from Mexico and 3 per cent or less from Malaysia, Indonesia and the Philippines. That means approximately 67 per cent of projects are in three of the four 'BRIC' (Brazil, Russia, India, China) countries, which are seen as emergent economies that will become major contributors to global growth in the 21st century (Kedia et al. 2006). China, which is the 'super-BRIC' in terms of its scale and growth rate, is also the largest recipient in the CDM programme. Approximately 84 per cent of the total annual CERs comes from China while another 11 per cent comes from India (UNFCCC 2011). This imbalance has affected the participation of many other developing countries in CDM policy development and raises issues about the legitimacy of CDM policy. On the demand side, European countries dominate, as might be expected given that the EU ETS has the strongest state regulatory support of any trading schemes to date. Over 80 per cent of CDM market investors are from the United Kingdom and Northern Ireland (29.43 per cent), Switzerland (19.53 per cent), the Netherlands

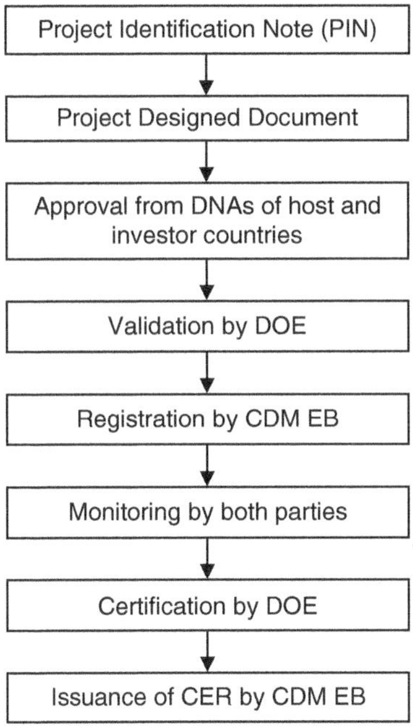

Figure 6.1 Project cycle for a CDM project

(10.85 per cent), Japan (11.52 per cent), Sweden (7.28 per cent) and Germany (5.31 per cent) (UNFCCC 2011).

The CDM approves projects through a complex and lengthy process relating to project design, method approval, validation, registration and verification, after which the project is issued with a CER offset (Lovell 2010: 354). The creation of a CDM project comprises eight basic stages as shown in Figure 6.1. At each stage, different organisations, with various roles, are involved. In order to participate in a CDM project, both the host (Annex B country) and the investor (Annex I country) must ratify the KP and establish a Designated National Authority (DNA). An Executive Board supervises the CDM (CDM EB) and accredits third parties known as Designated Operational Entities (DOEs) to independently validate project proposals and certify carbon credits. The CDM project must also pass an integrity test and meet additionality and permanence criteria. In addition, the reduction of GHGs has to be real, measurable and auditable. Additionality concerns the creation of extra benefits as a result of a given policy intervention above a specified baseline (Gillenwater 2012). In the case of the

CDM, this refers to 'real, measurable, and long-term benefits related to the mitigation of climate change... that are additional to any that would occur in the absence of the certified project activity' (UNFCCC 1998: 12).

The CDM project cycle starts with a Project Identification Note (PIN) that eventuates after a series of meetings between host and investor parties. This is followed by the writing of a Project Design Document (PDD) that discusses baseline scenarios, additionality conditions and project monitoring plans. This document is then submitted for approval by the DNAs of the countries involved and is validated by a DOE (Michaelowa 2007: 19). After receiving approval (validation) from a DOE, the project proposal can be submitted to the CDM EB. Before registration, the CDM EB sends the proposal for public comments. Once registered, the project participants are expected to monitor the project over its entire life. Finally, a DOE verifies and certifies GHG reductions, and the CDM EB issues CERs, or 'carbon credits'.

The buying and selling of credits require legal documentation, but once registered through the CDM Executive Board, carbon finance is channelled through the private sector or the various World Bank carbon funds that then finance the projects as they are implemented in the developing countries. Host countries and purchasers of credits must be in a country that is a Kyoto signatory. Host nations must provide a DNA to certify that the project contributes to sustainable development (part of the UNFCCC treaty requirements). Once operational the credits (CERs) are listed on the International Transaction Log (ITL) and may then be used by Annex I governments to reduce their emissions levels in compliance with their commitments under the Protocol. The ITL is designed to deliver transparent and efficient methods to account for CER trading and is one of the examples of the extremely complex regulatory mechanisms in the international compliance offset market (Lovell 2010).

Insofar as its institutional architecture permits, the CDM functions in a transparent and participatory manner. However, on the one hand, due to a lack of resources and financial and technical support from investors, many developing countries have fallen behind in capacity-building and related institutional development activities. This has led to claims of less favourable treatment towards some developing countries by investor parties and has been attributed partly to the low economic scale of projects and high transaction costs, as well as lack of clear policy that positively discriminates in favour of such countries. On the other hand, in the early 2000s, China benefitted considerably from several internationally funded CDM capacity-building projects and thus cornered the market opportunity presented by the CDM (Tulyasuwan 2008: 33).

The current procedures for CDM registration are also time-consuming. The methodology changes over time and participants are frequently asked to revise their approach once their 'request for registration' has been approved. Currently, there is about a 40 per cent chance that such a request will result in a review of approach. On average, the time lag between the request for registration and the actual registration of the project is around 150 days (IGES 2010). Project participants have consequently raised concerns over the capacity of the CDM EB to handle the level of activity. Two major reforms were discussed at COP 17 in Durban, including efforts to improve the CDM EB's work and establish an appeals process against CDM EB decisions (Climatico 2012: 12). The CDM programme has also been criticised in relation to the complexity and rigidity of procedures for accrediting projects (Brechet and Lussis 2006: 983; Sterk and Wittneben 2006: 274).

Implementation of the CDM and its effectiveness

In order to determine the effectiveness of the CDM and the validity of the arguments for and against, it is important to identify how the CDM's effectiveness is understood. In response to a given problem, states create initiatives designed to solve such problems by creating institutions or regimes. The question often arises as to why some international environmental agreements are more successful than others. While the concept of effectiveness can be simply defined in the sense that an environmental regime is successful if it is able to solve the problem for which it was created, such an approach may be unduly simplistic. Effectiveness is more precisely defined as a measure of the performance of a given institution in relation to a set of objectives (Elliott 2000). Policy analysts consequently tend to determine the success of environmental regimes by looking at the measures taken to implement the obligations agreed to, their outcomes (associated with changed behaviour) and impacts (results associated with changed behaviour, such as problem solving). Successful implementation has therefore been related to both behavioural and problem-solving effectiveness (Cadman 2011: 15). The following discussion examines, firstly, the degree to which the CDM has contributed to changing states' behaviour in terms of meeting the objective of sustainable development (an objective of the CDM), and secondly, the CDM's ability to contribute to solving the problem for which it was created (that is, the reduction of damaging GHG emissions).

Contribution to sustainable development

Case study one: unilateral CDM projects

Unilateral CDM projects allow a developing country to register a given initiative and sell the resulting CERs to the market. Acceptable sustainable development projects include the acquisition of advanced technologies, the promotion of employment, improving livelihoods and reducing the emissions of local pollutants. India is home to the most unilateral projects, with more than 87 per cent of their total investment (Seres 2008: 17). Such projects are developed by the host country and are not the concern of investors. However, developing countries are often left with only the most expensive project options available, as investors often already control the most lucrative low-cost CDM projects in-country (Karp and Liu 2001, p277; Muller 2007: 3205). Theoretically, as long as host countries are compensated for not just the immediate cost of a project but also the forgone option value, the problem will not arise, and host countries in fact have potentially much to gain from selling their low-cost projects. Host-country governments also have the option of not approving any CDM projects unless they get full compensation. However, it was also acknowledged in the early period of development of the unilateral option that a major reason for non-investment is the perceived risk levels in the particular developing country, and it was therefore unlikely that such countries would have the necessary capacity to develop such projects on their own, thus putting such countries at a further disadvantage when it comes to implementing sustainable development (Jahn et al. 2003). The development of local, small-scale projects is supposed to compensate for such disadvantage and reduce transaction costs through lower requirements for documentation, thus keeping CER-related costs down. However, financial flows for infrastructure development only occur once the CERs are sold, even though there has been early stage expenditure. Technology transfer has been identified as another important mechanism by which CDM projects may contribute to sustainable development. Although the CDM does not have an explicit technology transfer mandate, it has the potential to promote technology transfer by carefully selecting projects that need technology transfer from developed countries. However, due to the high transaction costs and perceived risks, industrialised countries are reluctant to invest in developing countries (Michaelowa 2007: 31).

Although 39 per cent of the total CDM projects are currently unilateral, they actually account for only about 21 per cent of total reductions attributed to the CDM (Seres 2008, p2.5). In the current CDM project pipeline, unilateral projects dominate, but their share often falls as they progress through the project cycle due to the financing problems identified above (Michaelowa 2007: 31). There are several studies on CDM projects and technology transfer (for details, see De Coninck et al. 2007; Dechezleprêtre et al. 2007; Haites et al. 2006; Seres 2008). Seres et al. (2009: 4922) undertook a comprehensive analysis, covering 3,296 registered and proposed projects. They reported that each occurrence of a unilateral project reduces the odds of technology transfer by 16 per cent. Unilateral projects attract little technology transfer, possibly also due to their smaller size. The rate of technology transfer is surprisingly low in India, which is the powerhouse of unilateral CDM projects (Seres 2008: 27). Developing countries can consequently end up as high-premium buyers rather than suppliers of an attractive investment option. The contribution of unilateral CDM projects to sustainable development, notably through technology transfer, is therefore somewhat questionable.

Case study two: HFC-23 and N_2O reduction projects

Studies show that HFC-23 reduction/avoidance and nitrous oxide (N_2O) reduction projects have little sustainability benefits (Liu 2008: 1873; Maosheng and Haites 2006: 160). HFC-23 is produced mainly as a waste product during the manufacture of another gas HFC-22 (for example R-22 or Genetron 22 or Freon 22). HFC-22 is used in some air conditioning systems and as a feedstock for high-performance plastics as a partial replacement for ozone-damaging gases (Wara and Victor 2008: 11). Similarly, N_2O is produced while producing adipic acid, a major constituent of nylon. N_2O and HFC-23 have 298 and 11,700 times more global warming potential than CO_2 respectively. Therefore, refrigerant companies find it cheaper to install an incinerator to burn the HFC-23, and once that is converted into CER, each tonne saved can be sold as 11,700 carbon credits. These companies are earning millions of Euros from these credits, more than from selling their refrigerator products. The cost of capturing and destroying HFC-23 at refrigerant plants is non-zero but very low (€0.5/tCO_2e) (Cosbey et al. 2005 cited in Liu 2008: 1873). Payments to refrigerant manufacturers, the Chinese government and to carbon market investors by governments and compliance buyers will in the end total ~€4.7 billion while estimated costs of abatement are likely less than €100 million (Wara and Victor 2008: 12).

There is a potential environmental problem created by this profit. The process of creating more refrigerants generates more HFC-23 to incinerate, thus increasing the net amount of pollution. Currently, there are 20 HFC projects in the world, and 11 of them are in China, where (as of 2007) over 90 per cent of the CERs issued stemmed from industrial HFC projects (CDM Market Brief 2007 cited in Liu 2008: 1873). The CDM Executive Board intervened to reduce the number of projects by 25 per cent, but by 2009 the total number of projects had been reduced by less than 10 per cent. Many investors from several countries (such as Canada, the Netherlands, Italy, Denmark, Finland, Sweden, Germany, the United Kingdom, Switzerland, Japan, Norway and Spain) have been involved in these HFC projects. It seems that there is general consensus among many parties to continue endorsing HFC projects.

In order to encourage other projects with a greater contribution to sustainable development outcomes, the Chinese enacted a royalty-sharing mechanism for all CDM projects. Projects with lower sustainable development benefits pay more to the government and vice versa (Liu 2008: 873; Maosheng and Haites 2006: 160). For fluorocarbon-related projects, 65 per cent of revenue goes to government and for N_2O projects 30 per cent. Energy efficiency improvements, the development and utilisation of new and renewable energy, methane recovery and utilisation, and afforestation and reforestation projects by contrast attract a levy of only 2 per cent (Maosheng and Haites 2006: 158). While this intervention is commendable, it has nevertheless meant that the CDM has become more focused on cost- and royalty-related issues, rather than concentrating on sustainable development.

Contribution to emissions reduction

As noted earlier, current CDM projects are highly concentrated in China, India, Brazil and Mexico (Table 6.1), with more than 70 per cent of the total expected annual CERs coming from China and India and those two countries receiving 72 per cent of the total US$95 billion investment (Seres 2008: 8). That means most of the investment is going to emerging countries that are more easily able to afford autonomous emissions reduction projects than is the case for the less developed countries – and increasingly so into the future. Although there are no formal inventories for China and India since 1994, there are several reports that claim these countries' GHG emissions rates have escalated. For example, Canadian economists Rubin and Tal (2008: 4) estimated that China's GHG emissions have increased by 120 per cent since 2000 while US emissions

Table 6.1 Statistics of registered CDM projects based on Project Designed Documents by 16 February 2010

Project types	Number of projects	Per cent of total No of projects	Total emission reductions by 2012 (1000 tCO_2e)	Per cent of emissions reduction by 2012	Total emissions reduction by 2020 (1000 tCO_2e)	Per cent of emissions reduction by 2020
HFC reduction and HFC avoidance	20	1	484,567	27	1,077,367	25
N_2O reduction	60	3	252,268	14	627,198	15
Biogas	292	14	64,122	4	127,401	3
Biomass	253	12	89,801	5	186,932	4
Methane recovery and utilisation	163	8	214,855	12	507,249	12
Hydro power	552	27	209,171	12	592,955	14
Fuel switch	54	3	110,751	6	312,380	7
Waste gas/heat utilisation	152	7	156,335	9	333,286	8
Wind power	283	14	126,562	7	322,551	8
Other	200	10	77,370	4	168,459	4

Source: Adapted from IJES (2010)

increased 16 per cent over the same period. Similarly, the Center for Global Development (2008) reported that as of 27 August 2008 China surpassed the United States as the world's biggest emitter of CO_2 from power generation. Furthermore, if the 27 Member States of the EU are counted as a single country, the EU would rank as the third biggest CO_2 polluter, after China and the United States. China's own report (NDRC 2007: 6) suggests that China's emissions increased by over 50 per cent from 1994 to 2004. Similarly, in India, total GHG emissions in 1998 were about 890 Million Tonnes CO_2 Equivalent ($MtCO_2e$) and 1,229 $MtCO_2e$ in 2006, an increase of 38 per cent (IGES 2009). These show that the contribution of CDM projects to the national GHG emissions reduction in China and India is insignificant.

The UNFCCC publishes inventory data for Annex I countries. On the basis of those data, Maraseni et al. (2010: 307–308) calculated the predicted changes in GHG emissions for Annex I countries with Land Use, Land Use Change and Forestry (LULUCF) activities between 1990 and 2010 (Table 6.2). Compared to 1990 levels, most of the Annex I countries

Table 6.2 Predicted changes in GHG emissions for Annex I countries with LULUCF activities between 1990 and 2010

KP countries	**Predicted change in GHG emissions between 1990 and 2010 (per cent)**
European Community	−2.60
Bulgaria	−57.60
Czech Republic	−26.80
Latvia	−160.50
Liechtenstein	23.30
Monaco	−3.10
Romania	−47.00
Slovakia	−34.10
US	21.10
Canada	62.20
Hungary	−34.90
Japan	5.20
Croatia	−21.40
New Zealand	17.90
Russian Federation	−42.30
Ukraine	−57.20
Norway	−18.70
Australia	5.20
Iceland	−7.50

Source: Adapted from Maraseni et al. (2010: 307–308)

are expected to have reduced their emissions by 2010, especially in the transition economies of the former Eastern Bloc and to a lesser extent in Western Europe. The Anglo-colonial countries, the United States, Canada, Australia and New Zealand are all expected to have increases. The net effect is, however, that between 1990 and 2010, the total GHG emissions of all Annex I countries are expected to reduce by 6.1 per cent, which is just enough to meet the collective Kyoto target of 5.2 per cent reduction during the first commitment period (Maraseni et al. 2010). China and India are expected to increase their GHG emissions significantly higher than the highest emitting Annex I countries by the first commitment period. If the current rates of GHG emissions in China and India are assumed to be constant, China's emissions since 1990 have increased by 100 per cent and India's emissions by 95 per cent. This raises serious issues regarding the legitimacy of CDM projects, as the two dominant CDM supplier countries (who should at least be reducing their emissions) are in fact increasing them at a large scale.

Conclusions and recommendations

In this chapter, different issues regarding the effectiveness of CDM implementation have been explored. There are arguments for and against the CDM. On the positive side, it is seen as an efficient way of tackling emissions reduction cost effectively, while promoting specialisation and North/South technology transfer and sustainable development that enhances local capabilities. On the negative side of the ledger, it is seen as a commodification of the global natural commons, converting it into private property for the purposes of market-based transactions. This reinforces domination hierarchies of the rich countries over the poor and alienates the poor from their land and decision-making over common resources. It also serves the developing countries as a means to avoid taking action themselves, while simultaneously advantaging rapidly developing countries' interests, such as China (Okereke 2010: 470).

The research demonstrated that the eligibility of HFC- and N_2O-related projects under the CDM have undermined the sustainable development objective of the mechanism. Much stricter criteria for HFC- and N_2O-related projects must be implemented to rule out those that have little sustainable development benefits. Similarly, the provision of the unilateral CDM project activity has benefitted the resource-advantaged countries rather than the intended least developed countries (LDCs), and it has reduced the chance of technology transfer. The rate of GHG emissions from countries that have higher number of CDM projects and have been issued higher number of CERs has been growing faster than the highest-emitting Annex I countries. These countries need to shoulder the responsibility of emissions reduction to a far more genuine level than at present.

References

Baumert, K., Kete, N. and Figueres, C. 2000. *Designing the clean development mechanism to meet the needs of a broad range of interests.* Washington, DC: World Resources Institute.

Brechet, T. and Lussis, B. 2006. The contribution of the clean development mechanism to national climate policies. *Journal of policy modeling,* 28, 981–994.

Cadman, T. 2011. Quality and legitimacy of global governance: case lessons from forestry. Basingstoke: Palgrave Macmillan.

Center for Global Development. 2008. *China passes U.S., leads world in power sector carbon emissions,* Center for Global Development, http://www.cgdev.org/, accessed 15 March 2010.

Climatico. 2012. *COP17 De-briefing: enhancement, decisions, and the Durban package.* Climatico Analysis, January 2012.

De Coninck, H.C., Haake, F. and van der Linden, N. 2007. Technology transfer in the clean development mechanism. *Climate policy,* 7 (5), 444–456.

Dechezleprêtre, A., Glachant, M. and Ménière, Y. 2007. The north-south transfer of climate-friendly technologies through the clean development mechanism. *Energy policy,* 36, 1273–1283.

Elliott, C. 2000. *Forest certification: a policy perspective.* Bogor: Center for International Forestry Research.

Gillenwater, M. 2012. *What is additionality? Part 1: a long-standing problem.* No location: GHG Management, ghginstitute.org/wp.../Additionality Paper_Part-1(ver3)FINAL.pdf, accessed 6 April 2012.

Haites, E., Duan, M. and Seres, S. 2006. Technology transfer by CDM projects. *Climate policy,* 6 (3), 327–344.

IGES (Institute for Global Environmental Strategies). 2010. *IGES CDM project data analysis,* updated 5 February 2012, http://www.iges.or.jp/en/index.html, accessed 26 February 2010.

IGES (Institute of Global Environmental Strategies). 2009. *CDM country fact sheet: India,* November 2009, http://www.iges.or.jp/en/index.html, accessed 16 February 2010.

IPCC (Intergovernmental Panel on Climate Change). 2007. *Climate change 2007: the physical science basis, summary for policy makers.* Cambridge: Cambridge University press.

Jahn, M., Michaelowa, A., Raubenheimer, S. and Liptow, H. 2003. *Climate protection programme unilateral CDM – chances and pitfalls.* Eschborn: Deutsche Gesellschaft für Technische Zusammenarbeit (GTZ).

Karp, L. and Liu, X. 2001. Clean development mechanism and its controversies. *In:* D.C. Hall and R.B. Howarth, eds. *The long-term economics of climate change: beyond a doubling of greenhouse gas concentrations.* Amsterdam: Elsevier Science B.V., 265–286.

Kedia, B., Lahiri, S. and Mukherjee, D. 2006. BRIC economies: earlier growth constraints, contemporary transformations and future potential, and key challenges. *In:* S. Jain, ed. *Emerging economies and the transformation of international business: Brazil, Russia, India, China (BRICS)*, Cheltenham: Edward Elgar, 46–74.

Liu, X. 2008. Rent extraction with a type-by-type scheme: an instrument to incorporate sustainable development into the CDM. *Energy policy*, 36, 1873–1878.

Lovell, H. 2010. Governing the carbon offset market, Wiley Interdisciplinary Reviews. *Climate change*, 1 (3), 353–362.

Maosheng, D. and Haites, E. 2006. Implementing the clean development mechanism in China. *International review for environmental strategies*, 6 (1), 153–168.

Maraseni, T.N., Maroulis, J. and Nooriafshar, M. 2010. Analysing the levels of human-induced greenhouse gas emissions from land use, land use change and forestry activities on Annex I countries ability to meet Kyoto targets. *International journal of environment and pollution*, 42 (4), 301–316.

Michaelowa, A. 2007. Unilateral CDM—can developing countries finance generation of greenhouse gas emission credits on their own? *International environmental agreements*, 7, 17–34.

Muller, A. 2007. How to make the clean development mechanism sustainable – the potential of rent extraction. *Energy policy*, 35 (6), 3203–3212.

NDRC (National Development and Reform Commission). 2007. *China's national climate change programme, prepared under the auspices of National Development and Reform Commission (NDRC) People's Republic of China*, June 2007, http://en.ndrc.gov.cn/newsrelease/P020070604561191006823.pdf, accessed 10 February 2010.

Okereke, C. 2010. Climate justice and the international regime, *Climate change*, 1 (3), 462–474.

Preston, B.L. and Jones, R.N. 2006. Climate change impacts on Australia and the benefits of early action to reduce global GHGs emissions. Australia: CSIRO.

Quiggin, J. and Horowitz, J. 2003. Costs of adjustment to climate change. *The Australian journal of agricultural and resource economics*, 47, 429–446.

Rubin, J. and Tal, B. 2008. *The carbon tariff*, http://research.cibcwm.com/economic_public/download/smar08.pdf, accessed 12 March 2008.

Seres, S., Haites, E. and Murphy, K. 2009. Analysis of technology transfer in CDM projects: An update. *Energy policy*, 37, 4919–4926.

Seres, S. 2008. *Analysis of technology transfer in CDM projects*. Bonn: UNFCCC Secretariat, 48p

Sterk, W. and Wittneben, B. 2006. Enhancing the clean development through sectoral approaches: definitions, applications and ways forward. *International environmental agreements*, 6, 271–287.

Teravainen, T. 2009. The challenge of sustainability in the politics of climate change: a Finnish perspective on the Clean Development Mechanism. *Politics*, 29 (3), 173–182

Tulyasuwan, N. 2008. The clean development mechanism in China – A win-win synergy? *A thesis submitted to the fulfillment of Bachelor Degree*. Lund University.

UNFCCC. 2011. *Clean development mechanism, the United Nations framework convention on climate change*. UNEP/WMO, http://cdm.unfccc.int/index.html, accessed 20 April 2011.

UNFCCC. 1998. *Kyoto protocol to the United Nations framework convention on climate change*. No location: United Nations. unfccc.int/resource/docs/convkp/kpeng.pdf, accessed 6 April 2012.

Wara, M. and Victor, D. 2008. *Realistic policy on international carbon offsets*. Published by program on energy and sustainable development. Working Paper #74 April 2008, Stanford University.

World Bank. 2009. *State and trends of the carbon market 2009*. Washington, DC, http://siteresources.worldbank.org, accessed 6 November 2009.

World Bank. 2008. *State and trends of the carbon market 2008*. Washington, DC, http://siteresources.worldbank.org/, accessed 16 May 2008.

7 Stakeholders in Climate Policy Instruments: What Role for Financial Institutions?

Matthew Haigh

Introduction

Everywhere the call is out for 'stakeholder' involvement as a means for improving developmental decisions, particularly those involving complex technology, uncertain risks and contending values. Everywhere but in funds management, it would seem. Despite the presence of obligations under policy instruments such as the Kyoto Protocol (KP), funds management sectors around the world (comprising pooled investment schemes such as hedge funds, pension funds, insurance companies and mutual funds) have been excluded from ecological crisis management discussions. Moreover, the interests of the ultimate beneficiaries of these fiduciary vehicles have not been factored in climate change discussions (Lohmann 2008: 362) and have not participated to any material extent in the mechanisms of the United Nations Framework Convention on Climate Control (UNFCCC).

The KP of the UNFCCC established three so-called flexible mechanisms for reducing GHG emissions: (i) GHG emissions trading systems located in some developed countries, including Europe and New Zealand, and at a regional level in North America (the Western Climate Initiative consisting of four Canadian provinces and six US states); (ii) the Clean Development Mechanism (CDM) and (iii) the Joint Implementation (JI). The CDM follows KP by allowing developing countries that are signatories to the Protocol (the so-called non-Annex I countries) to engage in sustainable development by means of GHG abatement projects. Developed (Annex I) countries may purchase the emissions saved (offset) by these projects in order to offset their own emissions and thereby meet their own committed KP-related emissions reduction target. The JI further allows developed countries to invest in developing country projects to offset their own domestic emissions, which can be also counted towards meeting their KP targets. The private funds under management that might be made available for environmental management efforts warrant attention on the extent to which financing measures designed for environmental management have both facilitated and precluded the involvement of pension schemes, insurance companies and mutual funds. This chapter examines the prevailing financing system of environmental pollution management and, on the basis of a series of interviews with investment-related stakeholders, discusses the institutional realities that impact the participation of the investment sector in financing environmental pollution management. A final section draws some conclusions and makes a number of policy recommendations.

Understanding stakeholder governance

A consensual 'new' governance that is understood in relational terms that contrast with earlier command/control governance models has emerged (for example Cadman 2011: 20, Fiorino 2010: 578, Foxon and Pearson 2008: 148). This multi-stakeholder view of governance is promoted in some policy circles and, to the extent that the global polity is involved, deserves examination for its applicability. This section looks at the development of a particular version of stakeholder governance relevant to fiduciary investors (as already mentioned, referring to pooled investment funds) that have claimed to be interested in climate and energy-usage policies, climate-related risks and other related issues.

The relational view of governance focuses on the necessary normative elements for collaborative social–political interaction. The relevant dynamic is said to involve a structure/process dialectic based upon participation and deliberation. Participation is understood in multidimensional terms of stakeholder accountability/transparency and inclusive interest representation. An effort is also made, in the interests of integrity, to retain social justice as a type of benchmark by referring to the need for equality and resources. Democracy is treated as a central component of the decision-making processes used in deliberation. Social justice and democratic and participative decision-making are treated as desirable characteristics of behavioural and problem-solving aspects of governance systems (Cadman 2011: 23).

According to this view, the governance of climate change abatement efforts becomes a matter of collaboration between and within

Table 7.1 Climate policy measures designed to attract financiers and investors

Research and technology development	Legislative and regulatory policies	Fiscal measures	Administrative measures
Permit trade	Standards	Carbon taxes	Labelling
Energy-usage subsidies	Carbon emissions and other environmental reporting	Feed-in tariff systems	Stakeholder forums
Clean development mechanism	Ambit of fiduciary obligation	Concessional tax status for green-certified investments	Disclosure of use of ecological considerations in the financing process
Joint implementation			Environmental audits

Note: See Lund (2007: 628–29) for a fuller explication of the connections and disconnections between environmental policy measures.

the institutional vehicles charged with addressing the relevant problems arising, including the UN regime and the World Bank as initiating and coordinating institutions. Administrative and facilitating programmes are the regime's Adaptation Fund, the Global Environment Facility, the KP (and national incarnations), the CDM and the JI, and attempts to create a global cap-and-trade system with appropriate allocation of tradable emissions quotas across countries and time, the most notable example being the EU carbon market (Table 7.1).

Certain connections and disconnections can be argued between some of these measures. Those that pertain to a fiduciary investor deserve consideration here. Pension funds in some European countries, Australia and New Zealand have been required to disclose their use of ecological considerations in the investment process (an administrative measure) (see, Haigh and Guthrie 2010). A fiscal measure would be a fiduciary's decision to allocate funds towards companies enjoying concessional tax status for green-certificated investments, such as has been made available in the Netherlands. Fiduciaries interested in forming assessments of the financial implications of climatic risks associated with companies in which they hold investments might find useful regulatory measures, such as mandated information disclosures of companies' carbon emissions levels, and research and development measures, such as the availability of energy-usage subsidies and utilisation of cap-and-trade allowances. Finally, if fiduciary duties were construed in trust law and fiduciary law as including consideration of ecological risks as they pertain to the portfolio, they might become regular issues for discussion at Board levels (Richardson 2011: 17). Present evidence suggests that the connections made above are latent; it is unlikely that most fiduciary investors would consider that climate policy measures as falling within their obligations (Richardson 2011: 5–10).

Coordination of the regime currently rests on solutions conceived in terms of least-cost mitigation and adaptation terms that do not permit easily the involvement of different political and economic actors (in particular, the private sector), nor allow for any significant degree of cooperation, decentralisation and competition (Zadek 2011: 1058). According to the constructivist view of governance considered here, the regime would seem to have permitted insufficient interest representation, limiting participation, while the rigid negotiating tactics used in international negotiations have not been particularly successful. Existing structures and processes have not facilitated collaborative interaction and consequently effective management of the economic, social and physical outcomes of climatic changes has not been achieved.

Despite the exposure of pooled investment funds to turbulent earth systems and despite the amounts of finance available in these sectors, private sources of finance have not been forthcoming (Andonova et al. 2009: 52; Zadek 2011: 1058). The markets' lack of interest in the regime is related to doubts about the credibility of the system. Various reasons have been put forward for markets' non-involvement in the policy regime. The wavering levels of commitment by signatories and the major exceptions to the regime in the cases of the United States and China have contributed to doubts about the likelihood of national governments and the major superpowers effectively tackling climate change (Fiorino 2010: 581). The delegation of procedural elements to national governments has caused permits to be distributed without charge, and quotas set at over-generous levels have made for a system of 'carbon credits' not attractive to investment. The absence of an apparent successor to the principal policy instrument and a lack of clarity between adaptation and mitigation efforts have left potential investors uncertain over the basic future architecture of the regime (Bäckstrand and Lövbrand 2007: 123). This has been combined with a series of unclear determinations as to how financing should operate (Haigh 2011: 1367). This has led to a lack of confidence that an appropriate risk-adjusted return will be attached to any of the regime's mechanisms, including cap-and-trade systems (Biermann and Gupta 2011: 1856; Marres 2011: 510). Pooled investment funds have therefore treated such elements of the regime as cap-and-trade, feed-in tariff systems and REDD+ as they might an exotic derivative, and have consequently not included them on their lists of permitted investments (Haigh and Guthrie 2010: 195–199). Moreover, nascent initiatives in financial investments in renewable energy projects are simply not at the scale at which institutional investors would consider (Bowen 2011: 1020). Investors may also be unwilling to engage in politically controversial asset classes like the CDM (for example Lohmann 2008).

It should be mentioned that there is no regulatory requirement for pooled investment funds to participate in cap-and-trade systems. Japan, the United Kingdom, New Zealand, Belgium, France, Germany and Denmark merely require managers of pooled investments to disclose 'the extent to which environmental considerations' are taken into account in portfolio construction (Haigh and Guthrie 2010: 147). Similarly, investor associations in the United Kingdom and Australia do not require investment teams to report on the validity of climate investments (Haigh and de Graaf 2009: 413–417). Market-driven regulation has been seen as contributing to the dissolution of responsibility for the negative externalities generated within the global economy (King and Lenox 2000: 698; Prakash and Potoski 2005: 350).

Pooled investment funds are rightfully looked at as major sources of finance yet are excluded from the regime's architecture. Callon's recognition of the European Emissions Trading System as a yet unfinished, experimental process of learning has value (Callon 2009: 548). But in so far as the involvement of stakeholders in opinion-, decision- and policymaking is concerned, the specification of stakeholders, rather than the financing, has not yet commenced. Participatory governance models, such as the one put forward by Cadman (2011), and evolutionary economic geography (Foxon and Pearson 2008) tend to assume that a transition to more climate-resilient and low-carbon economies is possible and moreover is path-dependent, meaning it has to somehow find a way through webs of institutions, knowledge networks, cultural values and complementary infrastructures. The method appealed to in the usual case is stakeholder participation. Robertson and Choi (2010: 90), in presenting their arguments for embedding governance considerations in ecological management, define organisational ecology as 'establishing structures and processes that facilitate collaborative dynamic among diverse participants, which, in turn, can enhance the quality of decisions made and implemented'. Robertson and Choi use this definition to identify the stakeholder connections between a triadic form of the global polity: the state; 'private', commercial sectors; and natural ecological systems. Cadman (2011: 29) too has argued that there is a need to connect resource providers, resource channels (legislators and policymakers) and resource recipients. Yet, it is plain that while the capital markets have been considered a latent policy resource for environmental pollution management (Lund 2007: 629), the communication channels between institutional investors and policymakers are dissipated (Haigh 2011: 1385, Robertson and Choi 2010: 99).

The realities of climate finance: what the investors think

Interview study

Interviews were used to provide insights into the ways investors have made sense of ecological crisis management, specifically their assessment of the UN regime. Thirty-two interviews were conducted in May through August of 2010 with senior executives in managed investment institutions, selected using professional networks of the author and according to the direction of relevant investor networks known to the author, for example, the Carbon Disclosure Project (based in Sydney, New York City and London). Organisations represented by the interviewees were located in the United States, Japan, Australia and Europe.

Interviewees were asked to describe how, if at all, they had participated in carbon-trading markets and if they had used data on energy-usage levels in their portfolio decisions. Interviewees were provided additional background material to clarify questions, if needed that typically took the form of a hypothetical investment decision-making scenario. One in three interviewees, at the time of interview, was a provider of research to portfolio managers. The next two largest groups were fiduciaries (28 per cent) and portfolio managers (22 per cent).

All interviewees doubted the stability of the regime architecture. A trustee of a European public-sector insurance company explained his reasons for refraining from permits trading:

> Carbon trading and all that kind of stuff: that's not an asset class that we are looking at very greedily. We don't understand it. We do not see a fit purpose from a risk perspective for us to go into that kind of asset class.

All interviewees questioned the operationalisation of the regime, and if they were 'expected' to participate. An environmental management analyst working in the London capital markets flagged his doubts on the credibility of the UN regime and pointed to the salience of a short-term perspective for investment processes.

> Doubts about the science have a real political impact. You factor in the time it will take to cost carbon, standardise information disclosures, and the right sort of regulation – that's five years. So at the moment, allocations according to carbon don't matter to most [people].

The interviewee quoted above explained that by 'five years' he/she was referring to the industry's common asset allocation planning horizon. Carbon reports were not considered useful for decision-making, a belief held by all interviewees. The reasons are to do with the regular work processes of investors that treat information expected to impact share price as relevant, as much as the informational quality issues associated with carbon reports. The following three interview extracts are illustrative.

1. Carbon emission data continue to be calculated and reported in different ways between regions, between companies, and sometimes even with companies. It's the wild west out there... (head of a 'responsible investment' advisory unit in a British insurance company)
2. The most critical policy in Australia is a price on carbon emissions. Reporting does not influence investment decisions in and of itself. Companies are generally careful to ensure that material price implications that may be reflected in their disclosures are addressed before they are disclosed [to data collectors and regulatory authorities]. This will change only once there is a price on carbon and [when] current disclosures attract a financial liability... (head of a Europe-based environmental lobby group comprised of institutional investors)
3. Information is not reliable because there is no standard for disclosure. It is difficult to understand materiality and relevance of information to price... (fiduciary of a North American stable of insurance companies and mutual fund trust structures)

The interview data obtained are suggestive of a conclusion that investors' participation in climate change management is contingent on institutional processes. In the financial markets, the subjectivity of participation in ecological crisis management differs from that of civic engagement, where the methods and techniques used in carbon accounting render public engagement visible (Marres 2011: 530). Portfolio managers, belaboured with client mandates, fiduciary obligation and subject to workplace conventions, are more likely to use carbon-accounting techniques that render their engagement invisible.

Discussion

Some consideration must be given to the relations between the real-life exigencies of pooled investment funds, and administration of the UN regime. With respect to fiduciary investors, there are very few 'devices of engagement' (Marres 2011: 517) that can help to enact a particular form of environmental participation as a legitimate means of investment practice. While policymakers have been eager to appropriate the discourse of financial services, they are yet to supply guidance on how policies on climate change, energy security, resource scarcity and so on might best be applied to wealth portfolios. Financial institutions have been left to arrive at determinations without the basic categories that usually accompany their decision-making.

Presenting as issues are the absence of a conceptualisation of transnational governance and a coordinating mechanism linking national regulatory systems. Participation in fiscal mechanisms by pooled investment funds might be frustrated given that central governments in the markets with which this chapter is concerned (with an exception in Norway) do not control the management contracts process of pooled funds (Clark and Hebb 2005: 2015). One might expect that uncertainties associated with derivative trading markets will be frowned upon by most pension funds, and acting as a conduit of firm-level measures such as energy-usage subsidies has not been in their usual remit. The interview extracts provided above suggest that fiduciaries have relegated the 'trading' elements of the UN regime at the level of a junk derivative. Certainly, the interviewees had no plans to participate in the major public financing mechanisms constructed to deal with the effects of global climatic changes.

Stakeholder dialogue designed to enable learning appears unlikely in the foreseeable future. A common issue for workers in the 'responsible investment' units of financial institutions in Europe and North America has been gaining access to and influencing investment teams and governing boards with their various agenda items (Haigh and de Graaf 2009). Influence pathways have not in the usual case relied on stakeholder representation but on broad-scale behavioural change (Haigh and de Graaf 2009: 409).

The effective exclusion of the capital markets from the climate policy regime is not the fault of policymakers nor of capital markets; it is to do with the interest representation permitted by a legally mandated, fiduciary duty to cater to and protect the economic interests of beneficiaries of vehicles like pension funds and insurance companies. Fiduciary and trust law require that for trustees and governing bodies of fiduciary investment vehicles, the beneficiaries of investment schemes are the relevant and only stakeholders. When a proposal arrives for a remit wider than the interests of immediate beneficiaries, such as a need for deeper and broader forms of climate financing, legal support is left wanting.

These remarks permit only a tentative interpretation of the beneficial impacts of the current programme of the management of climate change effects based on trading in GHG emissions allowances and rights linked to the creation of industrial projects. The calculation of the economic mitigation potential of emissions trading has been said to involve expected social benefits net of social costs, assuming that existing market barriers are removed by public mitigation policies. The conceptualisation that this entails is at the level of the individual, industrial firm; what would be required, however, is at the level of the global polity. Multilateral management of the effects of climate events depends on having a programmatic approach to ecological crisis management that links trading activities in financial derivatives to policy forums, in which financial and investment interests can participate meaningfully, rather than simply acting as the vehicles through which market systems operate.

Yield-maximising, risk-averse investors, based on the outcomes of the interview research presented above, have little incentive to adjust their operations unless required to do so by governments, informed by peer behaviour, or unless treated to economic prospects. The types of ecological crisis management mechanisms that can be considered for pooled investment funds include, following Lund (2007: 628–629), legislative and regulatory policies like standards and reporting; fiscal measures and research and technology development like permit trade, subsidies and taxation; and administrative measures like labelling and stakeholder forums. Available fiscal measures are those that relate to the income and cost levels of pooled investment funds, such as concessional tax status for green-certified investments (examples are provided by the Netherlands and China (Lund 2007: 628)). Relevant potential administrative instruments are a reconceptualisation of fiduciary obligation to include environmental and energy security risks, including the risk of ignoring energy insecurity, and requirement for pooled funds to disclose their exposure and responses to ecological crises to their beneficiaries (much like statutory financial disclosures).

The intermediated nature of capital market structure conditions investors' participation in measures such as permits trading and strategic investments in energy assets. Several reasons suggest themselves:

- It can be expected that institutional investors subject to fiduciary and trust law will pay short shrift to the values and concerns of those not represented immediately by beneficiaries of pension schemes and insurance companies.
- Commercial relationships imbue and constitute fiduciary capitalism, as the first duty of a pooled fund's board is delegation of investment management to experts. An intermediated structure operates against long-term financial assessment such as suggested by ecological considerations and, as much, 'stakeholder involvement'. Brokerages, custodians and advisory services used by fiduciaries, for example, are remunerated on the basis of the volumes of transactions.
- Further, while investment firms might be provided fiscal incentives to differentiate themselves from rivals and to signal their environmental commitment, differentiation might not represent a viable strategy for a well-diversified investor, as the portfolios of this type of investor are modelled on averaged benchmarks.

Connections between the ambit of fiduciary responsibility and environmental concerns are considered in the next and final section.

Conclusions and recommendations

Outcomes that can be associated with the UNFCCC infrastructure have been fewer than might have been hoped for. Extant initiatives (as at July 2011) are non-networked micro-financing at community levels and variously successful attempts of venture capitalists to launch renewable energy and clean technology financial products.

Unresolved issues have a tendency to turn into stubborn problems. In terms of climate justice, differencing relevancies can be linked to quietism on matters of access of the poor economies to basic services, for example, access to online weather alerts (Kym and Kouzmin 2009: 299–300). The reality of a future ecologically precarious, unjust world is given traction by multiple interlinked and systemic forces directing the policy pronouncements that have followed the KP. Governments have tended to make mention of the details of rapidly worsening living conditions in climate-vulnerable regions only so far as to reinforce a division of the rich West (the carbon mitigators) and the poor Other (the carbon adaptors).

Schnaiberg's (1980) treadmill of production theory is a widely recognised theoretical tradition in environmental sociology. Treadmill theory is based on a recognition of the dependence of societies on the natural environment, and the dramatic effects of contemporary societies on natural resources and ecosystems. How that might work out in finance capitalism involves interrelations wider than the state apparatus, financiers and companies. What is called for is the inter-subjective acknowledgement of responsibility, perhaps akin to what Cadman refers to as organisational responsibility (2011: 24). Only three of the thirty-two individuals interviewed for the research conducted for this chapter, when pressed, could articulate a sophisticated connection between turbulent earth systems and investment risk.

The oddness of the proposition that the funds management sector should participate in environmental financing (Clark and Hebb 2005: 2016) might, by challenging the contingent rationality of financial markets, create spaces for imaginings that will be needed for criticism and transformation. Approaches might be found useful to contextualise the notion of widened stakeholder governance examined above with the work-life exigencies of equity and bond markets. Stakeholder governance is (almost) precluded in the financial markets, with institutional processes such as networked contracts of intermediaries and service providers dispensing with democratic representation.

Recommendations for financial and intergovernmental policymakers

That the UN regime has not articulated with the funds under management in the capital markets suggests, if faintly, potential in networks of state and non-state actors (Andonova 2010: 25). Examples have appeared in the work of the Coalition for Environmentally Responsible Economies, the Carbon Disclosure Project and the Climate Disclosure Standards Board (CDSB), organisations that, independently of each other, have brought companies, investors and policymakers in North America, Europe, Asia and Australia together to discuss issues of environmental pollution management and without representation (that is, as non-partisan but connected individuals). Blatter (2009: 109) dispenses with representation requirements on grounds that not all stakeholders are experts, instead seeking a 'coalition' of people who 'share a set of normative and causal beliefs and who often act in concert'. More might be expected from meetings of policy experts, financiers, foresters, biologists, industrialists, philosophers and so on from a wide range of relevant disciplines making no claims to shared values (although their 'interests' might be shared), but a desire to learn something new. It should be noted that discussions on the efficacy of the KP measures have assumed that the market is an appropriate vehicle to bring about efficient and desirable policy outcomes. Reliance on the market is precarious, however, particularly when its regulation is handed to the main market protagonists (King and Lenox 2000: 700; Lohmann 2008: 360–264).

Fiduciary representation deserves consideration if the resources of managed capital are to be accessed for environmental management purposes. At issue is devising ways to widen the scope of interest representation in capital markets to include *all* the recipients of climate finance. The target beneficiaries of climate finance will almost certainly include stakeholders outside the fiduciary remit of trustees and governors. After all, not all those affected by climatic changes are members of pension funds and insurance policy holders. Networks of fiduciary investors (for example trustees), their service providers (for example 'asset' consultants and brokerage firms), climate scientists, environmental engineers and CEOs of businesses in climate-vulnerable regions might usefully be forged. However, more will be needed to make such networks effective in terms of policy outcomes. Conferences of the Parties (COPs) to the KP were designed to achieve the meeting of economic and environmental interests, disciplines and knowledge. The outcomes of recent COPs have not given investors reason to rush into carbon trading.

References

Andonova, L.B. 2010. Public-private partnerships for the earth: politics and patterns of hybrid authority in the multilateral system. *Global environmental politics*, 10 (2), 25–53.

Andonova, L.B., Betsill, M.M. and Bulkeley, H. 2009. Transnational climate governance. *Global environmental politics*, 9 (2), 52–73.

Bäckstrand, K. and Lövbrand, E. 2007. Climate governance beyond 2012: competing discourses of green governmentality, ecological modernisation and civic environmentalism. *In:* M.E. Pettenger, ed. *Global environmental governance.* Aldershot, UK: Ashgate, 123–149.

Biermann, F. and Gupta, A. 2011. Accountability and legitimacy in earth system governance: a research framework. *Ecological economics*, 70 (110), 1856–1864.

Blatter, J. 2009. Performing symbolic politics and international environmental regulation: tracing and theorising a causal mechanism beyond regime theory. *Global environmental politics*, 9 (4), 81–110.

Bowen, A. 2011. Raising climate finance to support developing country action: some economic considerations. *Climate policy*, 11 (3), 1020–1036.

Cadman, T. 2011. Evaluating the governance of responsible investment institutions: an environmental and social perspective. *Journal of sustainable finance and investment*, 1 (1), 20–29.

Callon, M. 2009. Civilising markets: carbon trading between in vitro and in vivo experiments. *Accounting, organizations and society*, 34 (3–4), 535–548.

Clark, G.L. and Hebb, T. 2005. Why should they care? the role of institutional investors in the market for corporate global responsibility. *Environment and planning A*, 37 (11), 2015–2031.

Fiorino, D.J. 2010. Sustainability as a conceptual focus for public administration. *Public administration review*, 70, S78–S88.

Foxon, T. and Pearson, P. 2008. Overcoming barriers to innovation and diffusion of cleaner technologies: some features of a sustainable innovation policy regime. *Journal of cleaner production*, 16(1), S148–S161.

Haigh, M. 2011. Climate policy and financial institutions. *Climate policy*, 11 (6), 1367–1385.

Haigh, M. and de Graaf, F.J. 2009. The implications of reform-oriented investment for regulation and governance. *Critical perspectives on accounting*, 20 (3), 319–417.

Haigh, M. and Guthrie, J. 2010. Management practices in Australasian ethical investment products: a role for regulation? *Business strategy and the environment*, 19 (3), 147–215.

King, A.A. and Lenox, M.J. 2000. Industry Self-regulation without sanctions: the chemical industry's responsible care program. *The academy of management journal*, 43 (4), 698–716.

Kym, T. and Kouzmin, A. 2009. Cyberpunk-web 1.0 'egoism' greets group-web 2.0 'narcissism': convergence, consumption, and surveillance in the digital divide. *Administrative theory and praxis*, 30 (3), 299–323.

Lohmann, L. 2008. Carbon trading, climate justice and the production of ignorance: ten examples. *Development*, 51, 359–365.

Lund, P.D. 2007. Effectiveness of policy measures in transforming the energy system. *Energy policy*, 35, 627–639.

Marres, N. 2011. The costs of public involvement: everyday devices of carbon accounting and the materialisation of participation. *Economy and society*, 40 (4), 510–533.

Prakash, A. and Potoski, M. 2005. Racing to the bottom? trade, environmental governance, and ISO 14001. *American journal of political science*, 50 (2), 350–364.

Richardson, B.J. 2011. From fiduciary duties to fiduciary relationships for socially responsible investing: responding to the will of beneficiaries. *Journal of Sustainable Finance & Investment*, 1 (1), 5–19.

Robertson, P.J. and Choi, T. 2010. Ecological governance: organizing principles for an emerging era. *Public administration review*, Special Issue, S89–S99.
Schnaiberg, A. 1980. *The environment: from surplus to scarcity*. New York, NY: Oxford University Press.
Zadek, S. 2011. Beyond climate finance: from accountability to productivity in addressing the climate challenge. *Climate policy*, 11 (3), 1058–1068.

8
Challenges for Global Health Governance in Responding to the Impacts of Climate Change on Human Health

Jeff Gow

Introduction

Human services protect, maintain or enhance the welfare or social functioning of individuals, families and social groups. These human services take many forms but primarily occur in the health, education and welfare domains. The links between climate change and human health are now well established (IPCC 2007: 8). Climate change is already contributing to the global burden of disease and premature death, and these effects are likely to increase in all countries (Hanna and Spickett 2011: 11S). The nature and extent of future health impacts will depend on the magnitude and nature of climate change (Maloney and Forbes 2010: 51). The risks to health of climate change will be incremental, increasing the burden of already occurring diseases like malaria, dengue fever and other major killers like diarrhoea and malnutrition. Poverty also makes people more vulnerable to many of the health effects of climate change, mainly through an inadequate access to health care (TIWGCCH 2010: 3). Poverty also increases the risk that a population already displaced by extreme weather events or environmental degradation will not easily recover and, as a result, will suffer much higher disease risks. Climate change impacts on health will also be unevenly distributed spatially. Densely populated coastal areas in small islands and developing states and mountain and arid zones will be particularly vulnerable.

Global health governance is characterised by a multiplicity of interested stakeholders who actively pursue diffuse, non-linear and, very often, non-aligned agendas. Across different diseases and illnesses a heterogeneity of approaches are evidenced in the global health response. Interested stakeholders include national governments of varying power, ranging from the United States to failed states like Somalia; intergovernmental organisations (IGOs) at the multilateral level like the United Nations and its many agencies and regional organisations like the African Union and the European Union (EU); public–private partnerships (PPPs) like the Global Fund to Fight AIDS, Tuberculosis and Malaria and the International AIDS Vaccine Initiative; and non-state actors such as philanthropic foundations like the Gates Foundation, non-government organisations like Doctors Without Borders and multinational corporations – especially drug companies.

The primary focus of this chapter is on the governance of global health service provision and how climate change will impact on the health of individuals and increase the burden of disease. Global health governance prior to the 1990s exhibited coherence, with the United Nations and its agencies at the forefront of effective and internationally coordinated responses to existing and emerging health threats. This broke down as the number of players interested in global health increased dramatically, many with self-interest as the central motivating factor for their involvement. Following Cadman (2011), an examination of the existing state of global health governance will be undertaken in relation to diseases and illness that will be made worse by climate change, with an emphasis on the criteria of interest representation, accountability and transparency (organisational responsibility), decision-making and implementation. Global health is currently characterised not by a single regime but by a complex of overlapping and competing governance arrangements across a range of illnesses and diseases. The existing complexities of global health governance and the magnitude and width of the threats posed to human health by climate change strongly suggest that many players will respond through the prism of their own self-interest and that no single governance regime will succeed. The response of global health governance to the impacts of climate change will consequently be fractured.

Impacts of climate change on human health

The health impacts of climate change are likely to vary from place to place and over time. Uncertainties about climatic variations, environmental changes and non-climate factors such as levels of education and health care in a country, its economic development and public health initiatives and infrastructure readiness mean that the precise effects of climate change on human health are difficult to discern and to quantify (IPCC 2007: 17). Most scientists, however, believe that the effects of climate change on health will be negative overall, and especially severe in developing countries (ibid: 19).

The health impacts of climate change include the direct impacts of extreme events, such as storms, floods, heat waves and fires, and the indirect effects of longer-term changes, such as drought, changes to the food and water supply, resource conflicts and population shifts (Rao 2009: 4). Increased vector-borne diseases may be associated with both direct and indirect effects (AMA 2008: 3). Some adverse impacts are already evident, while others are not expected to be felt until the middle of this century or even later. The World Health Organisation (WHO) has estimated that climate change is already responsible for measurable increases in malaria, diarrhoea and malnutrition (McMichael et al. 2004: 1544).

Projected climate changes are likely to affect the health status of millions of people, particularly those with low adaptive capacity. Some believe that the largest challenge posed by global climate change is likely to be the response to displaced populations or 'environmental refugees' (Patz and Khaliq 2002: 2284; see also Chapter 13, in this volume).

In the literature, seven categories of climate-related impacts and the potential effects these can have on health and well-being have been identified. *Temperature-related morbidity and mortality* – illness related to extreme cold and heat events, respiratory and cardiovascular illnesses and increased occupational health risks; *Weather-related natural hazards* – damaged public health infrastructure, injuries and illnesses, social and mental stress, population displacement; *Air quality* – increased exposure to outdoor and indoor air pollutants and allergens, respiratory diseases, heart attacks and other cardiovascular diseases, cancer; *Water- and food-borne contamination* – intestinal disorders and illnesses caused by chemical and biological contaminants; *Vector-borne and zoonotic diseases* – changed patterns of diseases caused by bacteria, viruses and other pathogens carried by mosquitoes, ticks and animals (Health Canada 2012: 2; IPCC 2007: 17; Rao 2009: 17–18). Malaria is the world's most widespread and fatal vector-borne disease, killing 1–2 million persons a year, the majority of these young children (Patz et al. 2007: 398). In fact, an estimated 25 per cent of all-cause mortality in children aged 0–4 years is directly attributed to malaria. Malaria transmission is highly influenced by climate and is one reason why sub-Saharan Africa is strongly affected by this killer disease (Rao 2009: 19). *Health effects of exposure to ultraviolet rays* – skin damage and skin cancer, cataracts, disturbed immune function (Rao 2009: 20). *Socio-economic impacts on community health and well-being* – the frequency, intensity and duration of extreme weather events can lead to increased demand for health-care services, disruption of social networks, interference with livelihoods, damage to or unavailability of adequate housing and shelter, damage to critical infrastructures (Health Canada 2012: 3).

Burden of climate change diseases

Climate change is expected to continue to contribute to the global burden of disease and premature deaths, especially in developing countries. For example, food security will be adversely affected as subsistence agriculture is expected to suffer, with Africa predicted to be severely affected as early as 2020, by which time water stress may affect 250 million people and crop yields in some countries are predicted to fall by 50 per cent (Chan 2007: 2).

The health effects of climate change are substantially diverse and regional differences in vulnerability to climate-sensitive diseases are significant, for example, from differences in climate exposures, public infrastructure/adaptability or baseline climate-sensitive disease rates (Patz et al. 2007: 399). Examples include regions that experience strong El Niño events (for example western South America, Southeast Asia and Africa) or locations experiencing concurrent environmental degradation that could modify (up or down) the strength of climate exposures (Patz and Olson 2006: 536).

A WHO estimate in 2000 of the global burden of disease attributable to climate change demonstrated that the African region had the greatest disease burden per million population (Table 8.1).

Table 8.1 Estimated health impacts of climate change by region measured in disability-adjusted life years (DALYs)

Region	Total DALYs	DALYs per million population
Africa	1,894,000	3,071
Eastern Europe & Russia	768,000	1,586
South America and Caribbean	92,000	188
South East Asia	2,572,000	1,703
Western Pacific	169,000	111
Developed Countries	8,000	9
World	5,517,000	920

Source: WHO (2003: 136)

The WHO analysis included only those health impacts that were underpinned by the strongest evidence – food- and water-borne disease, vector-borne disease, fatal injuries resulting from climate disasters and the risk of malnutrition – and is therefore likely to be an underestimate of the true burden of climate change on health. The disease burden estimate for the African region was 3,071 disability-adjusted life years (DALYs) per million, a figure that represents the sum of years of life lost due to premature death and years of life lived with disability.

The diseases/illnesses of malaria, tuberculosis, cholera, avian influenza, dengue fever, yellow fever and Lyme disease are those that are most likely to increase in prevalence due to climate change (TIWGCCH 2010: 4). This will be particularly true in those equatorial countries of Asia and Africa where these diseases are most prevalent and the impacts of climate change will be hardest felt. The high rates of co-infection of tuberculosis and Human Immunodeficiency Virus Infection/Acquired Immunodeficiency Syndrome (HIV/AIDS), especially in Africa, make the 33 million HIV positive people on that continent extremely vulnerable to climate change. Those vulnerable to these two diseases are also those through their location, mainly in Africa and Asia, most likely to also contract malaria. There are recorded somewhere between 350 and 500 million cases of malaria each year, with outbreaks dependent upon temperature, rainfall and humidity – all of which will be impacted by climate change (Rao 2009: 16). The Intergovernmental Panel on Climate Change (IPCC) expects the area of malaria transmission to increase in size as global temperatures increase, resulting in an additional 200 million cases per year (IPCC 2007: 19).

Global health governance and climate change

The following analysis focuses on the governance arrangements of interest representation, accountability and transparency (or organisational responsibility), decision-making and implementation as they are expressed within global health institutions and those diseases/illnesses that impact most of those most vulnerable to climate change and whose health will be made worse by climate change. It concentrates on the diseases/illnesses of malaria, tuberculosis and HIV/AIDS.

Interest representation

Global health is governed not by a single regime but by a complex regime of overlapping, competing and non-hierarchical regimes (Fidler 2010: 2). This complex is unlike single regimes that capture all players and decision-makers under one roof, such as the International Atomic Energy Agency (IAEA). It is a central intergovernmental agency for scientific and technical cooperation in the nuclear field, for the worldwide peaceful application of nuclear technology, application of safeguards and verification measures, providing international safeguards against the misuse of nuclear technology and materials, the promotion of nuclear safety and nuclear security standards and their implementation.

Competing and social and political interests have produced a multitude of governance systems in global health. HIV/AIDS provides an example of the proliferation of competing interests. After its discovery in the early 1980s, the WHO initiated a worldwide programme on AIDS. This was followed in 1994 by a resolution of the UN Economic and Social Council to launch the joint United Nations Programme on HIV and AIDS (UNAIDS) in 1996 as a multi-UN agency vehicle (nine) and the World Bank to mount and support an expanded response to HIV/AIDS, one that attempted to engage the efforts of many sectors and partners from government and civil society.

Yet UNAIDS is but one player in the global health effort against HIV/AIDS. Given the multiplicity of effects of HIV/AIDS it is framed much wider than purely as a health issue. Its effects in primarily poor countries exacerbate the already severe economic, social and developmental issues that places like Zimbabwe, Haiti or Kenya already confront. This has resulted in framing HIV/AIDS in multiple governance systems, which include the World Bank, WTO, G8 initiatives, regional efforts, bilateral programmes and new initiatives. The Global Fund to Fight AIDS, Tuberculosis and Malaria is an international financing institution that has committed US$22.6 billion in 150 countries over the past ten years to support large-scale prevention, treatment and care programmes against the three diseases. The HIV/AIDS example is a case study of how global health is governed by a multitude of interests, including states, IGOs, PPPs and non-state actors as outlined in Table 8.2.

The different players in global health have differentiated interests and hence their responses to the health challenges also differ. Many states implement programmes or provide aid assistance to others to advance their domestic interests rather than addressing the health needs of the

Table 8.2 Global health players in HIV/AIDS, tuberculosis and malaria

Player category	Examples	
States	Great powers	US, China
	Emerging powers	India, Brazil
	Developed states	Britain, Canada, Germany, Japan
	Developing countries	Bangladesh, Indonesia, Venezuela
	Failing or failed states	Congo, Haiti, Zimbabwe, Somalia
IGOs	Multilateral	UN (and its agencies), UNAIDS, WHO, World Bank, IMF, WTO
	Regional	African Union, ASEAN, European Union
PPPs	Mechanisms to increase access to health technologies	GAVI Alliance; Global Fund for HIV/AIDS, Malaria and Tuberculosis
	Drug and vaccine development partnerships	International AIDS Vaccine Initiative, Medicines for Malaria Venture, Malaria Vaccine Initiative, TB Alliance
Non-state actors	Philanthropic foundations	Clinton Foundation, Gates Foundation, Rockefeller Foundation
	NGOs and civil society groups	Amnesty International, Doctors Without Borders, Human Rights Watch, Oxfam
	Multinational corporations	Food and beverage, pharmaceutical and tobacco companies

Source: Adapted from Fidler (2010: 10)

recipients as revealed by the recipients themselves. The regional IGOs suffer from a similar problem in that the interests of the African Union states differ dramatically from the EU when the health effects of climate change are discussed. Those players without obvious self-interest motives, like the multilateral IGOs, most PPPs and most non-state actors except for multinational corporations, are best placed to advance the interests and needs of those adversely affected by climate change. Yet there is very little explicit incorporation of climate change effects into the governance of global health interests, nor is there any coordination of these interests through any representative body like a UN agency that could potentially coordinate responses.

Accountability and transparency

The various players in global health governance are not accountable to one central organisation for the success or otherwise of their activities.

In many instances, they are not even accountable to those they are meant to be helping. The programmatic activities of states, especially those great and developed states and their activities in global health, are driven by the internal needs and demands of domestic interest groups rather than those whose health is most affected by climate change. This is most starkly evidenced in the case of the prevention programme activities funded through the President's Emergency Plan for AIDS Relief (PEPFAR) in its initial iteration. A focus on sexual abstinence was a key plank of its prevention activities and funding for the provision of condoms was explicitly not allowed. So the most effective prevention method for the transmission of HIV (condoms) was explicitly not allowed to be funded, while programmes of dubious effectiveness like abstinence were funded. This situation arose from the effective efforts of fundamentalist Christian church lobby groups in Washington who opposed the science-based approaches of public health experts (Gow 2009: 5). The opposition to condoms in favour of abstinence was based on ideology rather than any independent assessment of the relative cost-effectiveness of the respective programmes. The implementation of PEPFAR-funded prevention activities in the affected countries had to follow the dictates of the US Government rather than be based on public health expertise or the desires and wishes of those countries receiving the prevention monies (Gow 2009: 7).

Given the lack of interest from global health players in climate change and its health impacts, it is unsurprising that accountability is lacking. The lack of collaboration and coordination among the multiple players in climate change health impacts is a critical problem. Given the different agendas of the multiplicity of actors ranging from indifference to ignorance, it is hardly surprising that the climate change response has been characterised by fragmentation, and where efforts have been made, duplication. Furthermore, transparency is insufficient, both with respect to IGOs and states (Gostin and Mok 2009: 15). Transparency supports accountability and both are closely associated with good governance. Unfortunately both characteristics are lacking in global health generally and climate change impact and responses specifically.

Decision-making

The much-reduced role of the United Nations and its agencies in global health governance means that some individual players/actors are implementing their own self-interested agendas. Often these individual actors make decisions without reference to those who will require or request health-care assistance. Great and developed states are particularly poor in this respect. With their large resources they implement health programmes and provide large levels of financial assistance on the basis of advancing their own self-interest rather than the health needs of recipients. The best/worst example of this scenario in global health in recent times has been the role of the US Government and its then leader, President George W. Bush, in unilaterally announcing the launch of PEPFAR in 16 countries without prior detailed discussion with the leaders of these mainly poor African countries (PEPFAR 2012: 1). PEPFAR was a commitment of US$15 billion over five years (2003–2008) from the United States to fight the global HIV/AIDS epidemic. The programme initially aimed to provide antiretroviral treatment (ART) to 2 million HIV-infected people in resource-limited settings, to prevent 7 million new infections, and to support and care for 10 million people (the '2–7–10 goals') by 2010. PEPFAR is the cornerstone and largest component of the US President's Global Health Initiative (GHI). With a special focus on improving the health of women, newborns and children, the GHI's goal is to save the greatest number of lives by increasing and building upon what works and, then, supporting countries as they work to improve the health of their own people (USAID 2012: 1–2). Nearly all of the diseases, conditions and illnesses that are the target of the GHI are closely linked to the human health changes induced by climate change, such as malaria.

PEPFAR and the GHI are examples where the most powerful and richest state in the world has unilaterally decided that the broader governance regime of international global health cooperation through international organisations like the WHO should be superseded. PEPFAR is the biggest and most notable unilateral action in global health governance history. This was a precursor of the GHI, in which the US government is pursuing a comprehensive whole-of-government approach to global health. The Initiative has funds of $63 billion over 6 years and has the dual objectives of achieving significant health improvements and supporting effective, efficient and country-led platforms for the sustainable delivery of essential health care and public health programmes. While these aims are admirable, they are being undertaken in isolation from the many other states, IGOs and non-state actors involved in global health. An explicit incorporation of the effects of climate change impacts is lacking in this instance and in the vast majority of GHIs regardless of the status of the actors involved, whether they be great powers, IGOs like the many UN agencies or many of the other powerful players. The major problem is that the individual players in global health governance have not explicitly incorporated climate change and its impacts into their activities.

As global health expert Laurie A. Garrett (2009) observed in connection with climate change:

> Despite a vigorous campaign to raise health concerns in advance of the [Copenhagen] summit, human disease and traumatic injury threats appear to have played no significant role in the debates, and there are only passing, insignificant references to them in the final, dismal document... It is impossible to imagine any ecological shift likely over the next fifty years that will more dramatically impact human health. Yet the 'health community' failed politically in Copenhagen.

Implementation

The global health governance regime is made up of several overlapping governance systems in which multiple players address specific problems (for example HIV/AIDS, malaria, tuberculosis) through different processes and by applying various principles (Fidler 2010: 4). The heterogeneity of each of these illnesses or diseases with their very different characteristics means that effective public health initiatives to address the threats are tailored specifically to each situation. A good example is HIV/AIDS. Methods for identifying the main source of infection through surveillance and invoking initiatives to reduce transmission or treat those already infected vary widely from region to region, country to country and continent to continent. Transmission pathways in more developed countries like the United States are mainly through men who have sex with men (MSM), whereas in sub-Saharan Africa (the epicentre of the epidemic) transmission is almost exclusively through heterosexual intercourse. In other areas, transmission is mainly through (unscreened) blood (China) and intravenous drug use (Russia). Effective mechanisms to reduce transmission are incredibly complex even within countries and regions with relatively homogeneous epidemics. Hence, programmatic initiatives that work in one place will be completely ineffective in another, even within small geographical areas. Further complicating matters is that if something does work in one disease and in one area, it does not mean that it will work for another disease even in the same area. A good example is the high rates of co-infection of HIV positive individuals in sub-Saharan Africa with tuberculosis. Surveillance patterns vary widely between the two diseases, as does the effectiveness of treatment, even in the same region. Although this link is well known and public health initiatives have attempted to combine surveillance and treatment responses and interventions, in many instances these have not been successful due to the heterogeneity problem. Thus, global health governance needs different regimes purposely built for specific challenges. Any type of regime is completely lacking in the context of climate change and human health.

Conclusions and recommendations

The current complex regime of global health governance with states, IGOs, PPPs and non-state actors all competing with each other results in sub-optimal outcomes for individuals and population health. New initiatives like the Global Fund and PEPFAR have further dissipated the chances for a consolidation of existing governance systems into a single coherent regime for managing HIV/AIDS or global health generally. The current regime of global health governance can be understood as comprising transnational and national actors, pursuing their own self-interest under a model of international cooperation (Ruger 2012: 655), and as such lack the degree of institutional legitimacy implied by Cadman in the Introduction. These actors have differing agendas and a selective set of initiatives that the HIV/AIDS and malaria examples indicate (Gostin and Mok 2009: 9). These are examples where dramatic increases in resources have occurred with little coordination. Further, these newly arrived global health players with their vast resources and/or political power have bypassed the traditional centralised global health response that IGOs like the WHO coordinated and implemented in the past in response to other global health challenges such as smallpox. Such outcomes are repeated across individual diseases and illnesses and will also be the case in the foreseeable future when discussions of the impacts of climate change on health arise. The multiplicity of impacts and their interconnectedness across social, economic and political arenas mean that the shape of any future global governance regime for health and climate change will also be complex by necessity. These circumstances result in a fracturing of the response to the urgent health needs of those most affected by the diseases and illnesses associated with climate change. Agreements among political players about the existence and extent of climate change are highly contested. The disappointing Copenhagen and Durban negotiations on global climate change provide further evidence of a contested global governance. Given these circumstances, the chances of a single regime for determining global health governance of the effects of climate change currently appear extremely limited.

Recommendations

The UN model is not broken and the Global Fund example of the coordination of international funding for HIV/AIDS, malaria and tuberculosis programmes throughout the world provides some hope yet for a similar outcome in governance for the health effects of climate change.

- Efforts should be made to attempt to bring together a comprehensive and collaborative international organisation to manage and coordinate the health effects of climate change.
- Leadership, which is inclusive not unilateral, organised not ad hoc, is needed to enable a coherent comprehensive response to the climate change effects on human health to be formulated and implemented.
- Collaboration and coordination of the multiple players are critical to enable an effective response to occur.
- Priority setting in a comprehensive framework with all stakeholders involved in charting a common course is also necessary for the programmatic response to the health effects to be effective.

These recommendations are to some extent overlapping; however, a systematic response is necessary to address these issues in an adequate manner given the increasing impacts of climate change on human health that are expected in the future.

References

Australian Medical Association (AMA). 2008. *Climate change and human health*. at http://www.dea.org.au, accessed 31 March 2012.

Cadman, T. 2011. *Quality and legitimacy of global governance: case lessons from forestry*. Basingstoke and New York: Palgrave Macmillan.

Cooper, A., Kirton, J. and Schrecker, T. eds. 2007. *Governing global health: challenge, response, innovation*. Aldershot: Ashgate.

Fidler, D. 2010. *The challenges of global health governance*. Working paper. New York: Council on Foreign Relations.

Garrett, L. 2009. *Council on foreign relations global health update*. 23 December 2009: 4.

Gostin, L. and Mok, E. 2009 Grand challenges in global health governance. *British medical bulletin*, 90: 7–18.

Gow, J. 2009. The adequacy of policy responses to the treatment needs of South Africans living with HIV (1999–2008): a case study. *Journal of the international AIDS society*, 12: 37.

Hanna, E. and Spickett, J. 2011. Climate change and human health: building Australia's adaptation capacity. *Asia-Pacific journal of public health*, 23(2) (supplement): 7–13.

Health Canada. 2012. *Understanding the health effects of climate change*, http://www.hc-sc.gc.ca/ewh-semt/climat/impact/index-eng.php, accessed 25 January 2012.

IPCC. 2007. Summary for policymakers. *In:* M.L. Parry, O.F. Canziani, J.P. Palutikof, P.J. van der Linden and C.E. Hanson, eds. *Climate change 2007: impacts, adaptation and vulnerability. Contribution of working group II to the fourth assessment report of the Intergovernmental Panel on Climate Change.* Cambridge: Cambridge University Press, 7–22.

Maloney, S. and Forbes, C. 2010. What effect will a few degrees of climate change have on human heat balance? Implications for human activity. *International journal of biometeorology,* 53: 31–51.

McMichael, A., Campbell-Lendrum, D., Kovats. S., et al. 2004. Climate change. *In:* M. Ezzati, A.D. Lopez, A. Rodgers, C. Mathers, eds. *Comparative quantification of health risks: global and regional burden of disease due to selected major risk factors.* Geneva: World Health Organization, 1543–1650.

Patz, J.A., Gibbs, H.K., Foley, J.A., Rogers, J.V. and Smith, K.R. 2007. Climate change and global health: quantifying a growing ethical crisis. *Eco health,* 4: 397–405.

Patz, J.A. and Olson, S. 2006. Climate change and health: global to local influences on disease risk. *Annals of tropical medicine and parasitology,* 100(5–6): 535–549.

Patz, J.A. and Khaliq, M. 2002. Global climate change and health: challenges for future practitioners, *Journal of the American medical association,* 287 (17), 2284.

PEPFAR. 2012. *The U.S. President's Emergency Plan for AIDS Relief (PEPFAR),* p://www.pepfar.gov/about/index.htm.%20Accessed%20August%2011, accessed 15 March 2012.

Rao, M. 2009. Climate change is deadly – the health impacts of climate change. *In*: J. Griffiths, M. Rao, F. Adshead and A. Thorpe, eds. *The health practitioner's guide to climate change.* London: Earthscan.

Ruger, J. 2012. Global health governance as shared health governance. *Journal of epidemiology and community health,* 66 (7), 653–661.

The Interagency Working Group on Climate Change and Health (TIWGCCH). 2010. *A human health perspective on climate change.* Washington, DC: Environmental Health Perspectives and National Institute of Environmental Health Sciences.

United States International AID Organization (USAID). 2012. *The U.S. Government Global Health Initiative,* http://transition.usaid.gov/ghi/factsheet.htm, accessed 1 August 2012.

World Health Organization. 2003. *How much disease would climate change cause?* Geneva: World Health Organization.

World Health Organization. 2000. *Climate change and human health: impact and adaptation.* Geneva: World Health Organization.

9
Climate Change and Sustainable Water Management

Jamie Pittock

Introduction

Climate change is one of the greatest challenges facing the world community; however, it is not the only problem that threatens humanity's survival on Earth. One assessment proposes that a safe living space for humanity requires environmental stewardship that keeps our footprint within nine planetary boundaries, including for freshwater, biodiversity, land use, emission of greenhouse gases (GHGs) and other pollutants (Rockström et al. 2009: 472–475). National governments have established numerous multilateral environmental agreements (MEAs) to facilitate sustainable environmental management, including 45 of global geographic scope and having at least 72 signatory countries (Inomata 2008: 10).

As one example of the interaction between climate change governance and other global challenges, this chapter examines how international institutions for climate change governance contribute to and detract from efforts to better manage freshwater resources and biodiversity, and proposes a number of solutions for more integrated managements building on an assessment by Pittock (2010b: 343–373). Specifically the United Nations Framework Convention on Climate Change (UNFCCC) with the United Nations Convention on Biological Diversity (CBD) and Ramsar Convention on Wetlands are considered to draw conclusions on governance quality and legitimacy (Ramsar 2009a; UN 1992a; 1992b). The objectives, commencement date and membership of the conventions are shown in Table 9.1. The Ramsar Convention has a broad mandate since its definition of wetlands embraces all freshwater aquatic ecosystems as well as shallow marine wetlands (Ramsar 2009a).

Climate change, biodiversity and water management issues are intimately linked. Freshwater ecosystems are particularly biodiverse and

Table 9.1 Assessment of UNFCCC, CBD and Ramsar conventions' comparative and overarching governance performance. Framework after Cadman (2011), data interpreted from Pittock (2010b: 345–369)

Convention and indicators/processes	**UNFCCC**	**CBD**	**Ramsar**	**Overarching**
Governance system	Framework convention elaborated by protocol and decisions, a secretariat, subsidiary bodies and an annual conference of parties	Framework convention elaborated by protocols and decisions, a secretariat, subsidiary bodies and a biennial conference of parties	Convention elaborated by decisions, a secretariat, subsidiary bodies and a triennial conference of parties	United Nations: General Assembly oversight with policy coordination via the UN Secretariat and UN Environment Programme
Participatory structure:	Poor:	Poor:	Good:	Poor:
• Inclusiveness	• Universal membership	• Universal membership	• Near – universal membership	• Universal membership
• Equality	• Smaller/developing state engagement constrained by limited resources	• Smaller/developing state engagement constrained by limited resources	• Smaller/developing state engagement constrained by limited resources	• Smaller/developing state engagement constrained by limited resources
• Resources	• Non-state engagement limited by high politics	• Non-state engagement as formal observers	• Non-state engagement as participating observers	• Non-state engagement as formal observers
• Accountability	• Annex A countries accountable on reporting of emissions but otherwise limited	• Accountability mechanisms are few	• Accountability via an integrated system for target setting and reporting. No enforcement	• Little consideration of sustainability issues by the General Assembly and Security Council. Limited mandate constrains UNEP
• Transparency	• History of high politics and closed room deals limit transparency	• High-level decision-making, limited national targets and reporting mechanisms limit transparency	• Transparent communication and reporting mechanisms	• High-level decision-making, limited national targets and reporting mechanisms

Table 9.1 (Continued)

Convention and indicators/processes	UNFCCC	CBD	Ramsar	Overarching
Deliberative process:	Poor:	Moderate:	Moderate:	Poor:
• Democracy	• Agreement by consensus may hinder progress	• Agreement by consensus may hinder progress	• Voting is possible but decisions are normally by consensus	• Agreement by voting
• Agreement	• No substantive agreement reached to fully address climate change thus far	• Targets to conserve biodiversity are agreed	• Targets to conserve wetlands are agreed with specific national implications	• Some sustainability targets agreed. Others are not. The framework is inadequate
• Dispute settlement	• Limited dispute settlement provisions	• Limited dispute settlement provisions	• No dispute settlement provisions	• Dispute settlement provisions are rarely applied
• Behavioural change	• Limited ensuing behaviour change by states thus far	• Limited ensuing behaviour change by states thus far	• Some behaviour change by states	• Limited ensuing behaviour change by states thus far
• Problem solving	• The climate change problem is not solved	• Biodiversity continues to be lost	• Wetlands continue to degrade	• Cross-sectoral integration and sustainability remain as challenges
• Durability	• There are a number of suggestions for alternative or complementary processes to bypass the gridlock in the convention	• CBD is accepted as the primary mechanism and there are proposals for additional mechanisms to be added under its purview	• Ramsar is accepted as the primary wetlands conservation agreement, and has a delegation from the CBD	• The UN is the only overarching global institution

Internal interaction	Poor: Outsourced scientific advice to the four-yearly IPCC, which is poorly linked to annual internal processes. Bureaucratic procedures and limited progress	Moderate: While internal procedures are bureaucratic agreement has been reached on targets and cross-sectoral work	Good: Integrated process of setting targets at different scales linked to planning and reporting. Cross-sectoral issues are addressed, resulting (on paper) in well-integrated work	Poor: Bureaucratic procedures, completion between agencies and lack of institutional reform limit progress
Cross-sectoral interaction	Poor: Agreed Joint Liaison Group process with the CBD has resulted in little change. Approaches by Ramsar have been rebuffed	Moderate: Making efforts through the Biodiversity and Joint liaison groups. Engagement with climate issues. Delegation to Ramsar	Good: Making strong efforts to engage the climate and biodiversity issues, including accepting delegation from the CBD	Poor: General edicts from the General Assembly for better integration have not been followed up. UNEP efforts to coordinate are hampered by its limited mandate and influence
Substantive and behavioural outcomes	Poor: Action on a scale to address the threat of climate change has not been agreed or implemented	Poor: Action to address biodiversity loss has been agreed but there is little evidence of behavioural change	Moderate: Comprehensive action to conserve wetlands has been agreed, and while trends in the condition of wetlands are negative in some areas substantial action is evident, e.g. reservation of wetland reserves	Low: The need for substantial action to achieve sustainability has been recognised but resulting institutional reforms are inadequate to respond. Cross-sectoral integration has been poor, especially between climate change and biodiversity institutions
Resulting legitimacy	Poor	Moderate	Good	Poor

contain more threatened biota than other biomes (MA 2005: 26; Pittock et al. 2008: 30). Growing populations, increasing wealth and consumption of thirstier commodities are dramatically increasing global demand for water, leading to scarcity in many regions (CAoWMiA 2007: 57–60; WWAP 2012: 44–68). Climate change is now having an impact on hydrology, changing the timing of availability and the quantity and quality of water (Bates et al. 2008: 3–4; Min et al. 2011: 378–381; Pall et al. 2011: 382–385). Less well appreciated is the indirect impact of climate change policies in increasing water consumption, including through greater use of technologies such as afforestation, carbon capture and storage, and biofuels (de Fraiture et al. 2008: 67; Hightower and Pierce 2008: 285–286; Hussey and Pittock 2012; Pittock et al. 2008: 32). Ideally, implementation of environmental agreements would optimise conservation of the climate, water resources and biodiversity. The CBD Secretariat stated this bluntly (SCBD 2003: 9):

> There is a clear opportunity to implement mutually beneficial activities (policies and projects) that take advantage of the synergies between the United Nations Framework Convention on Climate Change and its Kyoto Protocol, the Convention on Biological Diversity and broader national development objectives. [...] there is a lack of coordination among the multilateral environmental agreements, specifically among the mitigation and adaptation activities undertaken by Parties to the UNFCCC and its Kyoto Protocol, and activities to conserve and sustainably manage ecosystems undertaken by Parties to the Convention on Biological Diversity.

Similarly, the Ramsar Secretary General advised (Ramsar 2008a: 39):

> For the next triennium the key challenge is to make sure that the various agendas on climate change and other environmental concerns are not undermining the conservation and sustainable use of wetlands.

Assessing the institutional links

Ideally the three agreements would be assessed to identify the institutions that produce the best improvements in the state of the environment (Young 1999: 16–17; Young 2008: 17). Sadly, the UNFCCC shows no signs of preventing dangerous climate change, nor have the CBD and Ramsar halted the loss of biodiversity or wetlands. Hence this chapter focuses on questions of design: institutions that redirect the behaviours of relevant actors (Young 1999: 19–20; Young 2008: 17) and the institutional quality and legitimacy of governance arrangements (Cadman 2011: 1–18). Integrative mechanisms that manage conflicts, enhance positive synergies and use limited resources more effectively are prerequisites for more effective outcomes across the climate change, biodiversity and water sectors. Treaties have key roles in maximising rational utility, bestowing authority, facilitating learning and defining roles. They are most effective when they promote compliance by setting standards, enmeshing states and shaping expectations (Young 1999: 260–265). These positive influences are undermined when agreements conflict, for instance, should the UNFCCC promote the establishment of forest monocultures for carbon sequestration or biofuel production on biodiverse landscapes (French 2005: 108; Wolfrum and Matz 2003: 81, 87).

Integration is considered 'central to the attainment of sustainable development' (French 2005: 54). Integration is usually viewed as bridging the dichotomy between environment and development rather than interplay between environmental institutions. Better integration of MEAs is similar to the challenge in national governments in overcoming policy incoherence. Ross and Dovers (2008: 245) concluded that optimal national environmental policy implementation occurs when institutions in different sectors and across scales are linked and draw strength from constitutional and legal mechanisms, coordinating mechanisms (horizontal and vertical), influential lead agencies, and advisory-consultative- and review bodies. In this context, policy integration can be defined as the 'connection of goals, functions and processes across institutions to increase coordination and effectiveness' (Pittock 2010b: 352; Ross and Dovers 2008: 246).

Application of integrative mechanisms by the UNFCCC

The UNFCCC is compared here to the CBD and Ramsar in terms of evidence of conflicts and positive synergies and application of mechanisms for integrated outcomes as a basis for drawing conclusions on governance quality and legitimacy.

Institutional status

The three conventions have different characteristics that may influence their ability and willingness to collaborate. The UNFCCC and CBD are framework conventions (Vellinga et al. 2002: 293–296), whereas

Ramsar has been regarded as more narrow in its wetlands mandate and 'nested' under the informal CBD biodiversity umbrella (Wolfrum and Matz 2003: 65). The UNFCCC could but has not delegated to Ramsar the responsibility to implement common measures, for instance carbon sequestration in wetlands. The CDB has done this for conservation of inland waters biodiversity through a Joint Work Plan (JWP, CBD and Ramsar 2007). Conflicts over implementation may be forcing the CBD Secretariat to focus inwardly. The conventions have different levels of resources, with the UNFCCC having staff and a core budget nearly six times larger than Ramsar, which may provide incentives or barriers to collaboration (Pittock 2010b: 345).

Conflicts and positive synergies

Comparison of the approaches, concepts, objectives and obligations shows that the implementation of the UNFCCC could be positively synergistic with the CBD and Ramsar. The key difference is the first-order emphasis in the CBD and Ramsar on implementing an ecosystem approach (CBD 2009; Ramsar 2005), whereas the UNFCCC's Article 2 commits only to 'allow[ing] ecosystems to adapt naturally to climate change' (UN 1992b). Consequently, the UNFCCC retains the option of trading off degradation of some ecosystems to achieve particular climate change mitigation and adaptation objectives.

Examples of conflict: hydropower and afforestation

Hydropower illustrates the conflicts between global climate change governance and the sustainable management of freshwater ecosystems. Many nations have targets for expansion of hydropower as part of their national climate change policies (Pittock 2010a: 445, 2011). The Kyoto Protocol's (KP) Clean Development Mechanism (CDM) grants carbon credits to registered hydropower projects in developing countries, providing additional project income. As on 6 June 2012 there were 2,330 hydro projects registered or seeking registration amounting to 27 per cent of all CDM projects (International Rivers 2012). Regions with the greatest as-yet-unexploited hydroelectric power potential are particularly biodiverse, and Pittock argues that the CDM's environmental safeguards are ineffective (Pittock 2010b: 359–361). The CDM guidelines and approved methodology for hydropower projects only require environmental impact assessment 'undertaken in accordance with the procedures as required by the host Party' (CDM Executive Board 2008: 19) and do not require adherence to the procedures of other MEAs that may be of a higher standard (CBD 2002; Ramsar 1999).

Another example is forest establishment, which can result in significant reductions of inflows into streams (Herron et al. 2002: 369; van Dijk and Keenan 2007: 1). Few CDM projects involve afforestation and reforestation thus far, but this type of project is expected to increase: the CDM guidelines and approved methodology for these projects have cursory provision for assessing impacts that are unlikely to identify water use impacts (CDM Executive Board 2010; UNFCCC 2010).

Integrative mechanisms

The existence and use of integrative mechanisms – horizontal interaction – to manage the conflicts and enhance positive synergies is now considered, based on a review by Pittock (2010b: 354–356, 361–369). As conventions are largely implemented at the national scale through their contracting parties, a major issue is the degree to which convention agreements direct national-scale integration – vertical interaction.

Constitutional and legal mechanisms

National strategies, policies and plans translate commitments in MEAs into mandates, resources and implementing activities by Member States. Articles in all three conventions require their members to develop national strategies, which could be an opportunity to harmonise implementation of the conventions. The UNFCCC's Article 4.1 requires climate change to be considered in all relevant sectoral policies. The CBD has urged contracting parties to engage other conventions in preparing their national biodiversity strategy. Neither have monitored nor reported on progress globally. The 2009–2015 Ramsar Strategic Plan directs national wetland policies to be 'in place alongside and integrated with other strategic and planning processes' (Ramsar 2008b: Strategy 1.3).

Reporting is an important way for treaties to signal to members the importance of particular measures by linking convention priorities to national implementation, benchmarking performance and publicly reporting on performance at the national and international scales. The Ramsar Convention links its strategic plan to national activities and uses public reporting to encourage better implementation by contracting parties. It requires reporting on relevant biodiversity and climate change response measures. By contrast, the UNFCCC and CBD strategies are poorly linked to reporting that is less regular and more loosely synthesised. The UNFCCC requires regular reporting of emissions by developed nations and has a rigorous peer review process; however, this is not extended to other aspects (for example adaptation) or other members of the convention (Pittock 2010b: 362). There have been calls for harmonisation and streamlining of MEA reporting (CBD 2010a; IGM 2001), and approaches suggested to do this, but only Ramsar has encouraged its members to use them (EMG Secretariat 2008; Phillips 2005; UNEP and IUCN 2008).

MEAs are complex and a strategic plan may enable resources to be focused on key issues and collaborations and help to integrate different convention processes. Ramsar's Strategic Plan serves this function. The UNFCCC does not have a strategic plan but does adopt short-term statements of intent over a one- to two-year timeframe, whereas the CBD's Strategic Plan is aspirational and disconnected from other convention processes, such as its many Programmes of Work (Pittock 2010b: 363).

Compliance and enforcement mechanisms in the UNFCCC and CBD are restricted to components of their daughter protocols. The Ramsar Convention is explicit in its reliance on moral suasion to promote implementation by its members and uses information in national reports to 'name, praise and shame' national governments to engender compliance (Pittock 2010b: 363).

The only evidence of active UNFCCC collaboration with the CBD is through the Joint Liaison Group (JLG) of the Rio conventions. By contrast, Ramsar and the CBD have each adopted many measures on climate change (CBD 2010a, 2010c; Ramsar 2008c). The 1996 designation of the Ramsar Convention 'as the CBD's lead implementation partner on wetland issues' (SCBD 2008), and the implementation of the CBD-Ramsar JWP, is an example of best-practice collaboration.

Coordinating mechanisms

The JLG was established in 2001 and links the three Rio Conventions on climate, biodiversity and desertification. In 2007, they identified adaptation to climate change and reduction of deforestation as the 'main areas for future collaboration' (UNFCCC 2007). The CBD proposed at COP 10 the development of Joint Work Programme between the three conventions (CBD 2010b), but this proposal failed (CBD 2010c; *ENB* 2010). These decisions are clearly inadequate for managing global climate, water and biodiversity challenges (EMG Secretariat 2008). A similar Biodiversity Liaison Group of six biodiversity-related conventions – including Ramsar and the CBD – was formed in 1999, but has performed little better (CBD 2005, 2007; EMG Secretariat 2008).

There are extensive examples of the CBD and Ramsar conventions working together, aiding harmonisation of definitions, guidance and implementation of the conventions at the national scale (Pittock 2010b: 364). The CBD and Ramsar have frequently cross-referenced institutions of the UNFCCC, such as in major resolutions on climate change from 2002 to 2008; however, this has not been reciprocated.

Close coordination among each convention's national administrative authorities or focal points, has been proposed to achieving positive synergies in planning, implementation and reporting (GC UNEP 2010a; 2010b). There is no evidence of UNFCCC and erratic statements from the CBD encouraging their focal points to work together despite consistent urging by the Ramsar Convention and the liaison groups (CBD 2010a).

Alignment of the programme cycles of conventions, particularly Conferences of the Parties (COPs), has been proposed (WCMC 2004). It is difficult to see why coordination of meeting timing would not improve collaboration since the substance of issues discussed at COPs are developed over many years by subsidiary bodies, so there is ample time to coordinate measures if the conventions were motivated to do so.

Lead agencies

In 1996, the CBD invited Ramsar to be its lead partner in implementing CBD activities related to wetlands. While there is obvious potential for similar arrangements between the UNFCCC and other conventions, such opportunities have not been seized. Enhanced access to funds has been identified as a potentially powerful incentive for the integration of implementation efforts but much depends on the rules binding international financing mechanisms like the Global Environment Facility (GEF) (Wolfrum and Matz 2003: 194–202) or other climate change adaptation funds.

Advisory, consultative and review bodies

Outside of the JLG there has been no regular communication between the UNFCCC's Subsidiary Body for Scientific and Technological Advice (SBSTA) and its Subsidiary Body for Implementation (SBI) with comparable bodies of Ramsar and the CBD. By contrast, the CBD's Subsidiary Body on Scientific, Technical and Technological Advice (SBSTTA) and Ramsar's Scientific and Technical Review Panel (STRP) regularly participate in the work and meetings of one another (Bergkamp and Orlando 1999; Ramsar 2009b).

In the Ramsar Convention, non-government organisations are formally recognised as participating observers and have roles that can be categorised as providing information, developing policies, implementing measures, compliance and enforcement and promoting environmental justice (Gemmill and Bamidele-Izu 2002: 77; Pittock 2010b: 368–369; Pritchard 2006: 4). The UNFCCC and CBD admit civil society observers but in practice restrict their engagement to fewer roles at greater distance, for instance, by frequent designation of contact group negotiations as informal and closed to observers. Greater involvement of civil society actors could aid representation and participation in the UNFCCC (Dombrowski 2010: 397).

This assessment has identified examples of the adoption of mechanisms for more effective integration among the conventions, such as the JWP between Ramsar and the CBD. These better practices have not been systematically adopted by the UNFCCC, raising the questions of why the UNFCCC has avoided closer integration and whether reforms within the existing institutional framework are adequate or more radical changes required.

The lack of collaboration by the UNFCCC with the water and biodiversity-related conventions is surprising as it needs all the help it can get to avoid dangerous climate change. Pittock (2010b: 369–370) argues that there are four reasons that could explain why UNFCCC institutions have little engagement with other conventions: (a) actors influencing the regime consider climate change as the most urgent environmental problem; (b) the complexity of climate change issues and UNFCCC processes drive an inward focus; (c) agreement is yet to be reached on the most fundamental issue under the UNFCCC's auspices; and (d) the UNFCCC is considered by many to be a treaty about development and the economy – 'high politics' (Young 2002: 118–120) – and prioritised ahead of 'low-politics' environmental agreements.

Governance quality and legitimacy

Cadman (2011: 3-6) argues that the governance system determines the participatory structures and deliberative processes of MEAs and that the interaction between these two elements determines the outcomes and resulting legitimacy of global governance. In this chapter, the performance of the parallel climate change, biological diversity and wetland agreements has been assessed and compared. These data are summarised in Cadman's model in Table 9.1, which has been elaborated with a greater emphasis on the quality of internal interaction, horizontally and vertically in scale. Further, this assessment began by arguing that sector-by-sector approaches to sustaining the Earth's environment would fail unless there was strategic coordination between regimes, such as those for climate change, biological diversity and wetlands conservation to maximise positive synergies and minimise conflicts. Consequently Cadman's model is also elaborated by assessing the quality of interaction that each regime is undertaking horizontally with other international regimes (see Table 9.1 and Figure 9.1).

The three regimes studied can be argued to sit within overarching global governance institutions in the UN system. Although the Ramsar Convention is not formally a UN institution it has sought to work within the UN system with varying degrees of acceptance (as described above). Consequently, the overarching system is itself assessed (Table 9.1, Figure 9.1).

This assessment argues that the Ramsar Convention's legitimacy with government and non-government stakeholders is relatively good, due in large part to the more participatory elements of its structure, collaborative interactions internally and with other environmental sectors and the resulting substantive outcomes, such as designation and management of wetland reserves and regional wetland conservation programmes. The UNFCCC and overarching UN system are regarded here as having a poor performance in these respects, while the CBD's performance receives a moderate ranking.

How real are these differences? This assessment may in part reflect the high politics and more elite governance systems of the UNFCCC and UN systems compared to the room that the Ramsar Convention may have to engage stakeholders as a less contentious institution with a greater ability to demonstrate tangible outcomes on the ground. If so this is a case for establishing institutions like Ramsar (or segmenting existing agreements) that are more focused on particular and achievable elements of the global environmental governance challenge. Yet many of the high-level governance structures of the Ramsar Convention are similar to those of the UNFCCC and CBD: they are treaties governed by their nation state members. The differences in governance quality are then largely due to the many cultural and behavioural differences among these institutions as detailed in this assessment. There are no legal impediments to, for instance, the UNFCCC having joint work plans with the other conventions. This suggests that reform within the existing institutions could greatly improve outcomes and legitimacy.

A key question then is where the impetus for reform will come from within the regimes. In the case of Ramsar, the key decisions cited in this assessment arose from a combination of: (a) proposals from non-government stakeholders embedded in convention processes, (b) experts serving on a nimble Scientific and Technical Review Panel, (c) an activist convention secretariat and (d) a small number of engaged

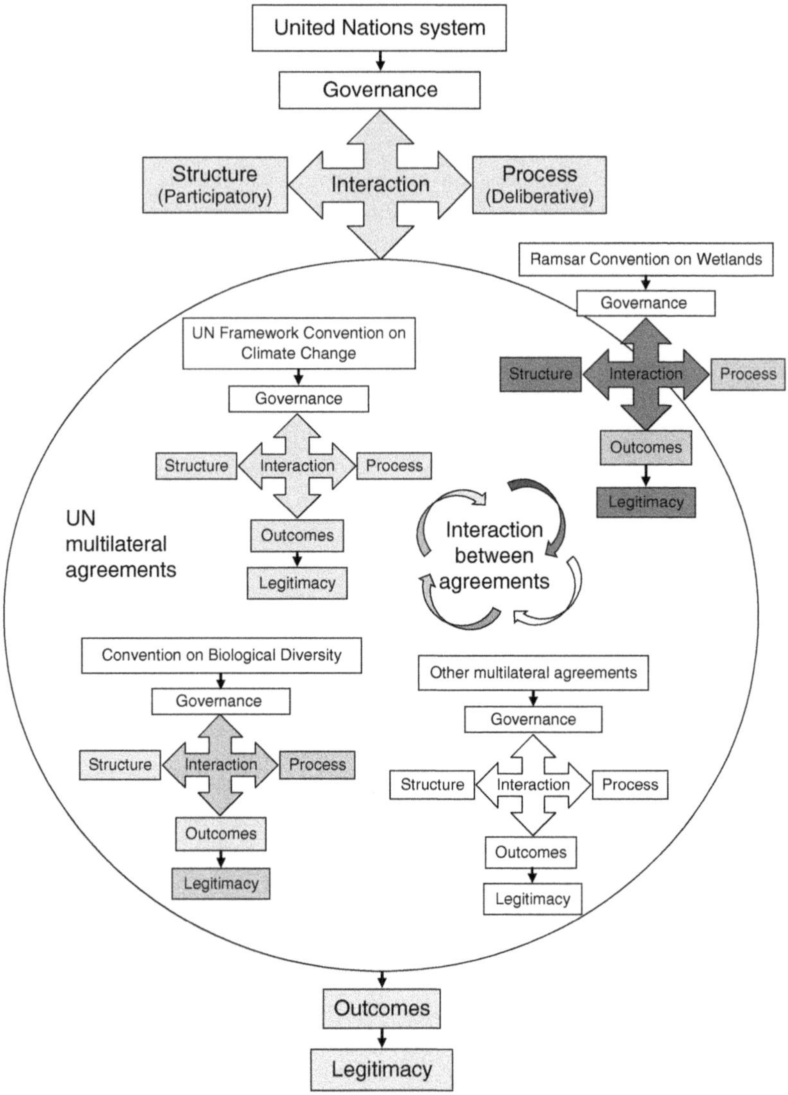

Figure 9.1 Governance quality and interaction: multilateral environment agreements

Notes: Dark grey = good institutional arrangements, mid grey = partly effective institutional arrangements, light grey = poor institutional arrangements, white = institutional arrangements not assessed. The Ramsar Convention is not formally within the United Nations' auspices.

Source: Adapted from Cadman (2011: 5) and Table 9.1

contracting party administrative authorities. Many of the more successful institutional elements of the Ramsar regime evolved from decisions taken at the 1996 Conference of Parties suggesting that it may take more than a decade to achieve substantially better governance quality.

Over the past decade, the options proposed for reforming global environmental governance have ranged from minimalist harmonisation of convention processes through to major institutional reform, such as the establishment of an overarching World Environment Organisation (Biermann et al. 2009: 351; France Diplomatie 2006). There is resistance to the establishment of new international institutions by many nations (GSP Secretariat 2011; WEF 2011) and a propensity to favour incremental reform (GC UNEP 2010a). This assessment suggests that global environmental governance could be greatly improved by application of the interactive measures described in this chapter. The obvious question is why such an approach would be more successful without substantial incentives given the lack of progress since the first of the Liaison Groups was established in 1999. Opportunities to better integrate the governance of conventions are unlikely to be realised without greater political will. In the short term, the most promising option may be additional funding restricted to collaborative implementation of the conventions at the national and international levels (Pittock 2010b: 371).

Conclusions and recommendations

Responding to climate change is a global priority. However, it is not the only planetary boundary. We need a global environmental governance system that maximises positive synergies among institutions responding to environmental priorities while minimising conflicts. This assessment of inter-sectoral environmental interaction shows that single-minded implementation of the UNFCCC is generating conflicts – as with hydropower development – and overlooking positive synergies with institutions for the conservation of freshwater resources and ecosystems. Better practices are identified, such as the exemplary JWP between Ramsar and the CBD.

The high politics of the UNFCCC and UN systems may detract from their ability to have more participatory structures and deliberative processes to enhance outcomes and legitimacy compared to more focused institutions like the Ramsar Convention. Yet the governance structures within the conventions are similar such that the differences in governance quality are largely due to cultural and behavioural differences.

This suggests that adoption of better institutional arrangements could lead to greater outcomes and increased governance legitimacy.

References

Bates, B.C., Kundzewicz, Z.W., Wu, S. and Palutikof, J.P., eds. 2008. *Climate change and water. Technical paper of the Intergovernmental Panel on Climate Change.* Geneva: IPCC Secretariat.

Bergkamp, G. and Orlando, B. 1999. *Wetlands and climate change. Exploring collaboration between the convention on wetlands (Ramsar, Iran, 1971) and the UN framework convention on climate change.* Gland: International Union for the Conservation of Nature.

Biermann, F., Davies, O. and van der Grijp, N. 2009. Environmental policy integration and the architecture of global environmental governance. *International environmental agreements: politics, law and economics*, 9 (4), 351–369. doi: 10.1007/s10784-009-9111-0

Cadman, T. 2011. *Quality and legitimacy of global governance: case lessons from forestry* Basingstoke: Palgrave Macmillan.

CAoWMiA. 2007. *Water for food, water for life. A comprehensive assessment of water management in agriculture.* London and Colombo: Earthscan & International Water Management Institute.

CBD. 2010a. *Cooperation with other conventions and international organisations and initiatives. Note by the executive secretary. UNEP/CBD/COP/10/17.* Montreal, QC: Convention on Biological Diversity.

CBD. 2010b. *Proposed elements for a joint work programme between the three Rio conventions on biodiversity, climate change and land degradation, subsidiary body on scientific, technical and technological advice, fourteenth meeting, Nairobi, 10–21 May 2010, Item 3.1.5 of the provisional agenda, UNEP/CBD/SBSTTA/14/6/Add.2.* Montreal, QC: Convention on Biological Diversity.

CBD. 2010c. X/33. Biodiversity and climate change. UNEP/CBD/COP/DEC/X/33 *Conference of the parties to the convention on biological diversity, tenth meeting, Nagoya, Japan, 18–29 October 2010.* Montreal, QC: Convention on Biological Diversity.

CBD. 2009. *Ecosystem approach.* Bonn: Convention on Biological Diversity. Available at: http://www.cbd.int/ecosystem/

CBD. 2007. *Options for cooperation.* Montreal, QC: Convention on Biological Diversity. Available at: http://www.cbd.int/cooperation/related-conventions/options.shtml

CBD & Ramsar. 2007. *Joint work plan (JWP) 2007-2010.* Montreal & Gland: Convention on Biological Diversity & Ramsar Convention on Wetlands.

CBD. 2005. *UNEP/CBD/WG-RI/1/7/Add.2 Ad hoc open-ended working group on review of implementation of the convention. First meeting Montreal, 5–9 September 2005. Item 5 of the provisional agenda. Cooperation with other conventions, organisations and initiatives, and engagement of stakeholders in the implementation of the convention.* Montreal, QC: Convention on Biological Diversity.

CBD. 2002. *COP 6 Decision VI/7 Identification, monitoring, indicators and assessments.* Montreal, QC: Convention on Biological Diversity.

CDM Executive Board. 2010. *Project design document form for afforestation and reforestation project activities (CDM-AR-PDD), Version 05*. Bonn: UN Framework Convention on Climate Change.

CDM Executive Board. 2008. *Guidelines for completing the project design document (CDM-PDD) and the proposed new baseline and monitoring methodologies (CDM-NM) (Version 7). EB 41 report annex 12*. Bonn: UN Framework Convention on Climate Change.

de Fraiture, C., Giordano, M. and Liao, Y. 2008. Biofuels and implications for agricultural water use: blue impacts of green energy. *Water policy*, 10 (S1), 67–81.

Dombrowski, K. 2010. Filling the gap? An analysis of non-governmental organizations responses to participation and representation deficits in global climate governance. *International environmental agreements: politics, law and economics*, 10, 397–416.

EMG Secretariat. 2008. *Enhancing MEA's coherence. EMG/IEG/MEAs/PD/29 Jan08*. Nairobi: Environmental Management Group Secretariat.

ENB 2010. Summary of the tenth conference of the parties to the Convention on Biological Diversity: 18-29 october 2010. *Earth negotiations bulletin*, 9 (544), [online].

France Diplomatie. 2006. *United Nations Environment Organization (UNEO)*. Paris: French Ministry of Foreign and European Affairs. Available at: http://www.diplomatie.gouv.fr/en/france-priorities_1/environment-sustainable-development_1097/united-nations-environment-organization-uneo_1966/index.html

French, D. 2005. *International law and policy of sustainable development*. Manchester: Manchester University Press.

GC UNEP. 2010a. UNEP/GC.26/3. Item 4 (c) of the provisional agenda. International environmental governance. Report of the Executive Director *Twenty-sixth session of the governing council/global ministerial environment forum, Nairobi, 21–24 February 2011*. Nairobi: Governing Council of the United Nations Environment Programme.

GC UNEP. 2010b. UNEP/GC.26/18. Item 4 (c) of the provisional agenda. Outcome of the work of the consultative group of ministers or high-level representatives on international environmental governance. Nairobi-Helsinki outcome, second meeting of the consultative group of ministers or high-level representatives on international environmental governance, Espoo, Finland, 21–23 November 2010 *Twenty-sixth session of the governing council/global ministerial environment forum, Nairobi, 21–24 February 2011*. Nairobi: Governing Council of the United Nations Environment Programme.

Gemmill, B. and Bamidele-Izu, A. 2002. The role of NGOs and civil society in global environmental governance. *In*: D.C. Esty and M.H. Ivanova, eds. *Global environmental governance options and opportunities*. New Haven, CT: Yale School of Forestry & Environmental Studies, 77–100.

GSP Secretariat. 2011. *High-Level panel on gloabl sustainability sherpas' meeting. Braunwald, Switzerland, 25 January 2011. Summary of discussion*. New York: UN Global Sustainability Panel Secretariat.

Herron, N., Davis, R. and Jones, R. 2002. The effects of large-scale afforestation and climate change on water allocation in the Macquarie River catchment, NSW, Australia. *Journal of environmental management*, 65 (4), 369–381.

Hightower, M. and Pierce, S.A. 2008. The energy challenge. *Nature*, 452 (7185), 285–286.

Hussey, K. and Pittock, J. 2012. The energy-water nexus: Managing the links between energy and water for a sustainable future. *Ecology and society*, 17 (1), 31 [online]. doi: http://dx.doi.org/10.5751/ES-04641-170131

IGM. 2001. *UNEP/IGM/2/5 Open-ended intergovernmental group of ministers or their representatives on international environmental governance. Second meeting Bonn, Germany, 17 July 2001. Proposal for a systematic approach to coordination of multilateral environmental agreements.* Nairobi: United Nations Environment Programme.

Inomata, T. 2008. *Management review of environmental governance within the United Nations system. JIU/REP/2008/3.* Geneva: United Nations Joint Inspection Unit.

International Rivers. 2012. *Spreadsheet of hydro projects in the CDM project pipeline.* Berkely: International Rivers.

MA (Millennium Assessment). 2005. *Ecosystems and human well-being: wetlands and water synthesis.* Washington, DC: World Resources Institute.

Min, S.-K., Zhang, X., Zwiers, F.W. and Hegerl, G.C. 2011. Human contribution to more-intense precipitation extremes. *Nature*, 470 (17 February 2011), 378–381.

Pall, P., Aina, T., Stone, D.A., Stott, P.A., Nozawa, T., Hilberts, A.G.J., Lohmann, D. and Allen, M.R., 2011. Anthropogenic greenhouse gas contribution to flood risk in England and Wales in autumn 2000. *Nature*, 470 (17 February 2011), 382–385.

Phillips, B. 2005. Mountains to the sea. *WWF Discussion paper.* Gland: WWF International.

Pittock, J. 2011. National climate change policies and sustainable water management: Conflicts and synergies. *Ecology and society*, 16 (2), 25. [online].

Pittock, J. 2010a. Better management of hydropower in an era of climate change. *Water alternatives*, 3 (2), 444–452.

Pittock, J. 2010b. A pale reflection of political reality: integration of global climate, wetland, and biodiversity agreements. *Climate law*, 1 (3), 343–373.

Pittock, J., Hansen, L.J., and Abell, R., 2008. Running dry: freshwater biodiversity, protected areas and climate change. *Biodiversity*, 9 (3 & 4), 30–38.

Pritchard, D. 2006. Ramsar's unique 'Partners' system keeps pace with the times. *MEA Bulletin* (15), 1, 4–5.

Ramsar. 2009a. *Convention on wetlands of international importance especially as waterfowl habitat. Ramsar (Iran), 2 February 1971. UN treaty series no. 14583. As amended by the Paris protocol, 3 December 1982, and Regina amendments, 28 May 1987.* Gland: Ramsar Convention on Wetlands.

Ramsar. 2009b. *How does Ramsar collaborate with other MEAs?* Gland: Ramsar Convention on Wetlands. Available at: http://www.ramsar.org/cda/en/ramsar-about-faqs-how-does-ramsar/main/ramsar/1-36-37%5E7723_4000_0__

Ramsar. 2008a. Agenda item XI. Ramsar COP10 DOC. 6. Report of the Secretary General on the implementation of the Convention at the global level *10th Meeting of the conference of the parties to the convention on wetlands (Ramsar, Iran, 1971).* Gland: Ramsar Convention on Wetlands.

Ramsar. 2008b. *Resolution X.1. The Ramsar strategic plan 2009–2015.* Gland: Ramsar Convention.

Ramsar. 2008c. *Resolution X.24 Climate change and wetlands.* Gland: Ramsar Convention.

Ramsar. 2005. *Resolution IX.1 Annex A: a conceptual framework for the wise use of wetlands and the maintenance of their ecological character.* Gland: Ramsar Convention on Wetlands.
Ramsar. 1999. *Resolution VII.16: The Ramsar convention and impact assessment: strategic, environmental and social.* Gland: Ramsar Convention on Wetlands.
Rockström, J., Steffen, W., Noone, K., Persson, A., Chapin, F.S., Lambin, E.F., Lenton, T.M., Scheffer, M., Folke, C., Schellnhuber, H.J., Nykvist, B., de Wit, C.A., Hughes, T., van der Leeuw, S., Rodhe, H., Sörlin, S., Snyder, P.K., Costanza, R., Svedin, U., Falkenmark, M., Karlberg, L., Corell, R.W., Fabry, V.J., Hansen, J., Walker, B., Liverman, D., Richardson, K., Crutzen, P. and Foley, J.A. 2009. A safe operating space for humanity. *Nature*, 461 (24 September 2009), 472–475.
Ross, A. and Dovers, S. 2008. Making the harder yards: Environmental policy integration in Australia. *Australian journal of public administration*, 67 (3), 245–260.
SCBD. 2008. *25 September 2008. Notification. Streamlining and making more explicit the roles of the CBD and Ramsar Convention Ref: SCBD/STTM/DCO/va/65040.* Montreal, QC: Secretariat of the Convention on Biological Diversity.
SCBD. 2003. *Interlinkages between biological diversity and climate change. Advice on the integration of biodiversity considerations into the implementation of the United Nations framework convention on climate change and its Kyoto protocol.* Montreal, QC: Secretariat of the Convention on Biological Diversity.
UN. 1992a. Convention on biological diversity (with annexes). Concluded at Rio de Janerio on 5 June 1992 *treaty series, No. 30619. Multilateral.* New York: United Nations.
UN. 1992b. *United Nations Framework Convention on Climate Change.* New York: United Nations.
UNEP and IUCN. 2008. *TEMATEA issue-based modules for coherent implementation of biodiversity-related conventions.* Nairobi and Gland: UNEP and IUCN.
UNFCCC. 2007. *Report of the seventh meeting of the joint liaison group of the convention on biological diversity, the United Nations framework convention on climate change and the United Nations convention to combat desertification. Bonn, 7 June 2007.* Bonn: United Nations Framework Convention on Climate Change.
UNFCCC – UN Framework Convention on Climate Change. 2010. *Draft text: ad hoc working group on long-term cooperative action under the convention twelfth session Tianjin, 4–9 October 2010,* UNFCCC/AWGLCA/2010/14. No location: United Nations.
van Dijk, A.I.J.M. and Keenan, R.J. 2007. Planted forests and water in perspective. *Forest ecology and management*, 251 (1–2), 1–9.
Vellinga, P., Howarth, R., and Gupta, J. 2002. Improving global environmental governance. *International environmental agreements: politics, law and economics*, 2, 293–296.
WCMC, 2004. *Synergies and cooperation. A status report on activities promoting synergies and cooperation between multilateral environmental agreements, in particular biodiversity-related conventions, and related mechanisms.* Cambridge: UNEP World Conservation Monitoring Centre.
WEF. 2011. *Rio+20 and the future of international environmental governance. Discussion paper – January, 2011.* Davos: World Economic Forum.
Wolfrum, R. and Matz, N. 2003. *Conflicts in international environmental law.* Berlin: Springer-Verlag.

WWAP (World Water Assessment Programme). 2012. *The United Nations world water development report 4: Managing water under uncertainty and risk.* Paris: UNESCO.

Young, O. 2008. Institutions and environmental change: The scientific legacy of a decade of IDGEC research. *In*: O. Young, L.A. King and H. Schroeder, eds. *Institutions and environmental change: principle findings, applications and research frontier*. Cambridge, MA and London: The MIT Press, 3–46.

Young, O. 2002. *The institutional dimensions of environmental change: fit, interplay and scale*. Boston, MA: MIT Press.

Young, O.R., ed. 1999. *The effectiveness of international environmental regimes: causal connections and behavioral mechanisms*. Cambridge, MA: The MIT Press.

10

Food Security, Food Sovereignty and Global Governance Regimes in the Context of Climate Change and Food Availability

Nick Rose

Introduction

Global governance, as Henk Overbeek notes, was an initially radical (and hence marginal) concept when it first emerged in the 1970s, challenging the then-prevailing configurations of political economic power via proposals for a New International Economic Order, to be presided over by 'humane and democratic institutions' (Overbeek et al. 2010: 697–698). Following the Soviet Union's collapse and the neo-liberal counter-revolution, global governance became domesticated and then mainstreamed into a 'euphemism for the global rule of capital' (Overbeek et al. 2010: 697). As neo-liberal market ideology falters in the aftermath of the tumultuous events of 2008–2009, Overbeek detects the early signs of a Polanyian 'double-movement', in the form of a turn from 'market-based governance to state-led government', and he concludes by quipping that 'the discussion about "global governance" [will soon become] a debate among historians rather than social scientists' (Overbeek et al. 2010: 701–702).

Whether such an assessment is well founded, premature, ill-conceived or overly simplistic remains to be seen, but Overbeek's main point – the need to historicise and politicise a concept that is frequently portrayed as apolitical and ahistorical – is well made. This need is especially pressing in the sphere of the global governance of food, where, in the past decade and a half, a major schism has opened up between those economic and political forces seeking to further 'constitutionalis[e] private property rights, guarante[e] the unhindered mobility of capital, and control and subordinat[e] potentially rebellious forces and states' (Overbeek 2004: 53 cited in Overbeek et al. 2010) and those counter social movement forces who, together with a growing number of states, are seeking to roll back such constitutionalisation.

The global governance of food has become, in other words, a highly contested and politicised arena. At the discursive and policy level, this contestation expresses itself in the form of a debate between, on the one hand, a marketised conceptualisation of *food security*, articulated in institutions such as the World Bank and the World Trade Organisation (WTO) and in informal gatherings such as the G8 and the G20 and, on the other, an increasingly anti-market, or at least, anti-free market, conceptualisation of *food sovereignty*, articulated by the global small farmer movement La Via Campesina and its supporters in various non-governmental organisations. At the material level, the contestation frequently takes the form of practices of – and resistance to – 'accumulation by dispossession' (Harvey 2003: 71–73, 139–145). This phrase captures a wide range of the more destructive aspects of contemporary capitalism, often linked to financial and currency crises (Klein 2007: 6–11). In the case of food and agriculture, it describes and explains phenomena such as the ongoing 'land-grabs', whereby indigenous and peasant communities are forced off their ancestral lands to make way for the 'industrialised grain-livestock complex' that has facilitated the exponential growth of the factory farming, fast food and agro-fuel sectors (GRAIN 2010: 139–153; Tokar 2010: 126; Weis 2007: 16–20). Alternatively, the food sovereignty vision calls for a multiplicity of diverse, participatory and democratically controlled local and regional food systems, planned and designed to serve the interests and well-being of farmers, communities and ecosystems (Wittman et al. 2010: 2–5).

The climate crisis and the food crisis thus form the backdrop of a 'theatre for "governance through markets" ' in which a 'new model of green governance' consists in corporate self-regulation, bolstered via government incentives and subsidies (Peet et al. 2010: 26–27). This governance model is being shaped at a time when the severity of the climate crisis and its potential to create a permanent and intensifying global food crisis is becoming difficult to ignore. Severe and extreme droughts, floods and storm events around the globe strongly suggest that food production in many of humanity's principal agricultural zones will be severely compromised as average global temperatures increase (Lynas 2009: 9, 185–187). Warming in excess of as little as one degree centigrade may produce perennial droughts in the western United States, which could leave half or more of the entire US corn crop in a poor or extremely poor condition. Four degrees of warming would devastate up to 40 per cent of China's basic grain crops and render grain production extremely difficult in much of India (ibid). With many poor countries food-import-dependent, there is an increased risk of significant crop failures that could result in higher food prices and rising levels of food insecurity.

The cost-externalisation of climate change-related impacts onto the poor has the potential to become 'the biggest market failure ever seen' (Stern 2008: 1). The market environmentalism of the post-Kyoto climate change regime assumes that the way to protect the environment – and, simultaneously, to feed the world – is 'to price nature's services, assign property rights, and trade these services within a global market' (Bumpus and Liverman 2010: 266). However, attempts to fashion a consensus around this policy framework have encountered the difficulty that the global food and agricultural system has been revealed to be highly dysfunctional (Magdoff and Tokar 2010: 14–26). Contemporary forms of food, fibre, timber and biofuel production are some of the largest drivers behind deforestation (Hassan et al. 2005: 243–269), which, together with its fossil-fuel-dependent nature, means that the globalised and industrialised food system accounts for between a quarter and a third of all greenhouse gas (GHG) emissions (Beddington et al. 2012: 12). Hence the critique that this food system is ultimately self-destructive, undermining the essential conditions for its own reproduction (Moore 2010: 398; Weis 2010: 316, 324, 328).

The originators of food sovereignty are the farmer activists of La Via Campesina, the global movement for small family and peasant farmers, landless rural workers, Indigenous peoples, women and youth, which embraces an estimated 200 million people affiliated to its 150 member organisations, located in 70 countries across 5 continents. According to Nettie Wiebe, former president of the Canadian National Farmers Union, one of the founding organisations of La Via Campesina, food sovereignty was collectively developed and first articulated by La Via Campesina in 1995, in explicit juxtaposition to the concept of 'food security' as it was then being deployed by core capitalist states in the lead-up to the World Food Summit in Rome in 1996 (interview with the author, 15 September, 2010). In governance terms, food sovereignty proposes the replacement of a globalised, oligarchic food system with democratised and localised food systems, leading to a 'fundamental shift in values expressed in changing social and political relations' (Wittman et al. 2010: 4). In concrete policy terms, the emphasis is on redistributive agrarian reform and associated measures to support smaller-scale farmers; the transition away from chemical and fossil fuel intensive agriculture to agro-ecology; and the diversification of food distribution systems through more localised food economies.

This chapter contrasts the prevailing food security governance regime with the emergent food sovereignty regime and asks whether the two are irreconcilably antagonistic or whether some accommodation might be possible. Of particular interest from the governance perspective is the development in the last few years of a reformed, inclusive and participatory Committee on World Food Security (CFS). Setting itself the ambitious task of achieving policy convergence and coherence in a fragmented and contested field, the CFS has embraced a vision of being 'the foremost inclusive international and intergovernmental platform for a broad range of committed stakeholders [who] strive for a world free from hunger [and] the progressive realization of the right to adequate food in the context of national food security' (CFS 2009: para. 4).

Global governance of food and agriculture

According to Eric Holt-Giménez and Annie Shattuck, there are two distinct trends in global food and agricultural governance that might be termed 'neoliberal' and 'reformist' (Holt-Giménez and Shattuck 2011: 135). The neo-liberal trend is 'managed by institutions such as the United States Department of Agriculture, the European Union's Common Agricultural Policy, the World Trade Organization, the private sector financing arm of the World Bank, and the IMF', in collaboration with 'the major agri-food monopolies, [the] agricultural policies of the G8 and big philanthropy capital', such as the Bill and Melinda Gates Foundation (Holt-Giménez and Shattuck 2011: 115, 119). The reformist trend is 'managed by weaker offices within the same institutions' as well as the principal food-focused agencies of the United Nations (UN): the Food and Agriculture Organisation (FAO), the World Food Programme, the International Fund for Agricultural Development (IFAD) (Holt-Giménez and Shattuck 2011: 121). In recent years, the UN human rights institutions, in particular the Special Rapporteurs on the Right to Food, have also been increasingly active in this space, with the normative development of the right to food bringing a sharper rights-based focus to interpretations of food security.

From the governance perspective, although the institutions of the neo-liberal trend have no specific mandate to manage food security, their decision-making processes have been setting the key policy parameters in which global and national-level responses are being implemented (Holt-Giménez and Shattuck 2011: 119, 121, 124). Conversely, the reformist institutions charged with the management of food security have struggled to balance the realities of the geo-political balance of power and the imperatives of capital accumulation, with an emerging scientific consensus that the globalised food system is in need of drastic reorientation.

The clearest expression of this consensus came in the form of the 2008 report of the International Assessment of Agricultural Knowledge, Science and Technology for Development (IAASTD 2008), an unprecedented inter-disciplinary project that brought together the work of hundreds of agricultural and development experts. The report urged a 'fundamental shift' in agri-food systems to prioritise the livelihood needs of small-scale farmers; redress the inequalities experienced by women in agriculture; and embrace and adopt the principles and methodologies of agro-ecology (IAASTD 2008: 4–7). The report rejected the idea that a 'business-as-usual' market-based approach would be at all sufficient to meet the food security challenges of the 21st century. It placed particular emphasis on the need to tackle the deep inequalities in the global political economy and to address the vulnerabilities faced by marginalised communities and the rural poor, especially in light of the medium-term challenges posed to food security by climate change (IAASTD 2008: 3–4). In these respects, this report constitutes a strong endorsement of some of the key policy shifts demanded by La Via Campesina and the food sovereignty movement. In their normative development of the right to food, the UN Special Rapporteurs have likewise supported the food sovereignty approach and have issued strongly worded denunciations of the marketised food security approach (de Schutter 2011).

During the same period and as a result of sustained pressure over several years from civil society actors and the Special Rapporteurs, the Member States of the FAO agreed to reform the CFS, elevating it to become the pre-eminent global governance institution with specific responsibility for global food and nutrition security, and substantially changing its rules of procedure, making it a genuinely inclusive and participatory governance forum. Explaining the rationale behind the need for a reformed CFS, the second UN Special Rapporteur on the right to food, Olivier de Schutter, set forth five key reasons why existing global food security governance had 'failed to eradicate hunger and severe malnutrition': 'an almost exclusive focus on increasing agricultural production, instead of the adoption of a more holistic view about the causes of food insecurity'; 'a failure of global governance to overcome existing fragmentation of efforts'; 'a still incomplete understanding of how to work in certain areas which have an impact on our ability to achieve food security for all'; 'a failure to follow upon commitments, itself a result of a lack of accountability'; and 'the insufficiency of national strategies for the realisation of the right to food at [the] domestic level' (de Schutter 2009: 2–3).

As regards fragmentation of existing governance arrangements, de Schutter noted how a multiplicity of the UN and other agencies and institutions offered 'conflicting advice to states' on matters pertaining to food security, and how the 'policy space' of states in this area had been constrained through the impact of neo-liberal structural adjustment policies. These shortcomings highlighted the need for a more coordinated, integrated and comprehensive approach to the governance of food and agriculture, with a focus on the key goal of combating hunger (de Schutter 2009: 2). In terms of failures of implementation, de Schutter noted the common practice of the making of solemn declarations of intent at various summits and conferences, but the lack of any follow through on commitments made, which he attributed to the absence of 'strong leadership from the highest level of government' to prioritise realisation of the right to food (de Schutter 2009: 2). This failure was compounded by the lack of 'accountability mechanisms at the national level to protect the right to food' (de Schutter 2009: 2).

In terms of its structure and process, the new CFS is 'composed of members, participants and observers and will seek to achieve a balance between inclusiveness and effectiveness' (CFS 2009: para. 7). The CFS Plenary (see Figure 10.1) is supported by a High-Level Panel of Experts (HLPE) charged with providing, first, analysis 'of the current state of food security and its underlying causes'; secondly, 'scientific and knowledge-based analysis and advice on specific policy-relevant issues'; and finally, the identification of 'emerging issues [to help] members prioritize future actions' (CFS 2009: para. 37).

As regards its deliberations and decision-making processes, the annual Plenary is the principal forum (CFS 2009: paras. 20–22). Ultimate decision-making power, including the drafting of final reports of Plenary sessions, is reserved to Member States. The Plenary is supported by the CFS Bureau, which ensures coordination among all actors and levels and undertakes intersessional tasks (CFS 2009: paras. 29–30). The Bureau is in turn supported by an Advisory Group that includes four members of the Civil Society Mechanism (CSM) (CFS 2009: paras. 31–32).

In its pre-eminent role of investigating the causes of food insecurity, the HLPE tackled the issue of climate change in its June 2012 report (HLPE 2012). This report contained several important conclusions

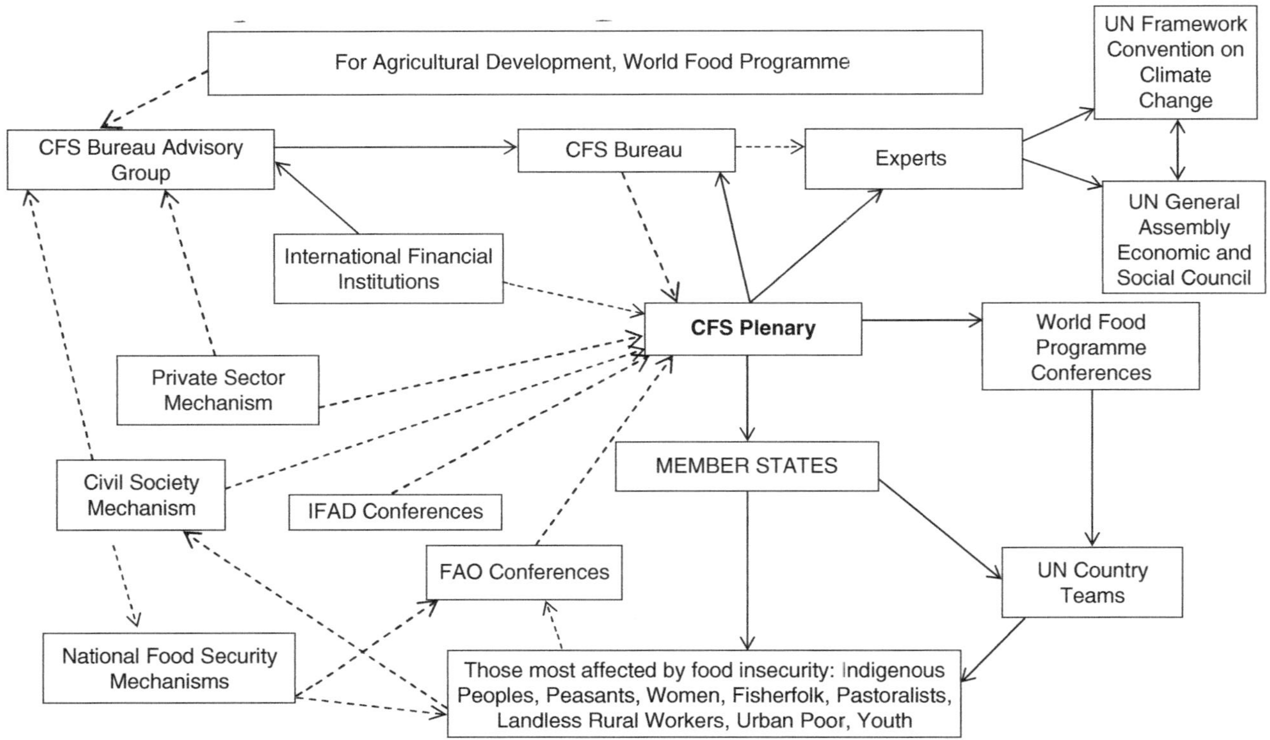

Figure 10.1 CFS organogram

Note: --→ Governance Inputs; ⟶ Governance Outputs.

Source: Adapted from Sobhan et al. (2012)

and recommendations, making explicit the many connections between climate change and food security. It urged Member States to develop their own climate change and food security mitigation and adaptation strategies, with a particular focus on addressing the needs of poor and vulnerable communities, such as those living in coastal areas. Among other recommendations, the HLPE also called for their recommendations on climate change and food security to be included in the World Food Summit (WFS) Global Strategic Framework (GSF), the first draft of which was released in October 2012, and for both the Intergovernmental Panel on Climate Change (IPCC) and the United Nations Framework Convention on Climate Change (UNFCCC) process to incorporate food security considerations and promote climate-resilient food systems in their future deliberations (ibid) (Figure 10.1).

Analysis of the global food governance regime(s)

This analysis is in three parts. First, there is a brief consideration of the (neo-liberal) preference for marketised food security. Next, there is a longer discussion of the reformed CFS. Finally, the emergent food sovereignty approach, led by La Via Campesina, is summarised, focusing on the proposed Declaration of the Rights of Peasants and the institutionalisation of food sovereignty in Ecuador as a concrete example of what a legislative and institutional framework for food sovereignty actually looks like in practice. The appropriateness of each model to meeting humanity's food needs for the 21st century in the context of climate change will also be briefly discussed.

Neo-liberal food security

Following Cadman's framework (2011: 15–18), a qualitative analysis indicates that the neo-liberal model of global food governance would rate poorly in terms of both meaningful participation and productive deliberation. In terms of interest representation, the main interests that are included are those of corporate agri-business. As noted above, some largely tokenistic attention has been paid in recent years to the needs of smallholder farmers and women, but neither they nor their representatives are included in any of the key institutional forums. Decision-making in these forums is notorious for its lack of democracy, with European and American governments exercising disproportionate power in both the IMF and the World Bank, and thus shaping outcomes favourable to their interests and those of transnational capital (Monbiot 2004: 16; Wilkinson 2011: 44).

Critics of the ostensibly consensus-based decision-making of the WTO argue that the key discussions take place through opaque processes, behind closed doors or in corridors, among the so-called quads – the governments of the European Union, the United States, Canada and Japan – and a selected few other 'key' countries (Bello 2000:18). These discussions are then presented as *fait accompli* (Monbiot 2004: 16–17, Steinberg 2002: 341).

In addition to substantive differences on the content of negotiations, not the least of which are the terms on which further liberalisation of trade in agriculture may take place (Downes 2006: 621), such non-transparent, non-inclusive and non-democratic governance arrangements have played their part in the stalling of the current Doha 'development round' of WTO talks for the best part of a decade (Downes 2006: 621). According to Nettie Wiebe, this does not mean that the 'free trade' agenda championed by the United States has been abandoned; rather, it is now being pursued outside the WTO, in bilateral and regional trade agreements, such as the Trans-Pacific Partnership Agreement (interview with the author, 15 September 2010). In its June 2012 report on climate change and food security, the HLPE urged governments to devise strategies to make food systems more sustainable and durable in the face of the bio-physical and socio-economic impacts of climate change. Notably, it called for governments to support 'food production and distribution practices that are more resource efficient *and* have fewer environmental externalities'. It remains to be seen if this recommendation will be met.

Reformed Committee on World Food Security

The reformed CFS seeks to address the governance failures identified by Olivier de Schutter via the embrace of 'three guiding principles – inclusiveness, strong linkages to the field to ensure the process is based on the reality on the ground, and flexibility in implementation' (CFS 2009: para. 3). The CFS is mandated with a series of governance roles to be implemented in two phases from October 2009. In the first phase, the roles of the CFS are those of global-level coordination, facilitating policy convergence and providing support and advice to countries and regions (CFS 2009: para. 5). In its second phase, the CFS will assume a greater degree of action at the national and regional levels, moving from a facilitation to a coordination role. Secondly, it will strengthen accountability mechanisms to hold states and other actors responsible for their commitments to eliminate hunger and ensure the right to food for all, including through the 'definition of common indicators to monitor progress'. Finally, the CFS will take the lead in the development of a GSF for food security and nutrition to 'improve coordination and guide synchronized action by a wide range of stakeholders' (CFS 2009: para. 6).

A most novel feature of the reformed CFS and one that speaks strongly towards more meaningful participation is the creation by civil society and people's organisations of their own autonomous CSM, which was endorsed by Member States at the 36th Plenary session of the CFS in October 2010. Together with other CFS participants, the CSM is accorded a number of rights of participation, which means that the CFS also performs better with regard to both inclusiveness and equality. It places particular emphasis on those organisations representing the most marginalised and vulnerable groups, especially women, as mandated by the CFS itself, and has acknowledged the need to provide 'significant global level resources in order to facilitate the development of capacities and processes at regional, sub-regional and national levels' (CFS 2009: para. 11(ii)).

The intention behind the CFS is to see that existing global rules and policy frameworks on trade, finance, investment and economic development are rendered coherent with the core human rights obligations of states and 'with human dignity, solidarity and people's sovereignty' (CSM 2010). Central to the task of achieving such coherence is the proposed GSF on food and nutrition security, a second draft of which will be presented to the 38th Plenary of the CFS in October 2012. As noted, the HLPE has also urged that the CFS incorporate climate change in its deliberations leading up to the GSF. The elaboration of the GSF is an ambitious undertaking. Equally ambitious are the efforts to 'secur[e] multi-stakeholder consensus across the full spectrum of CFS stakeholders' (CFS 2012: para. 3). Issues of sharp contention and division have already emerged, with the CSM working statement highlighting concerns over the definition of food sovereignty, women's and workers' rights, and 'very weak' provisions on monitoring and the accountability of States, intergovernmental institutions and transnational corporations (CSM, n.d.). Whether, and to what extent, these concerns are taken into account and addressed when the Member States in the Plenary of the CFS approve the final draft of the GSF will be indicative of the quality of the new food global governance regime that the CFS potentially represents.

The HLPE's report on climate change and food security reveals many synergies with the food sovereignty emphasis on protecting smaller-scale farmers and recognising the central role these farmers have to play in building climate-resilient food systems. The governance processes of the CFS facilitate decision-making that is informed, both by best-practice scientific advice in the form of the HLPE and by civil society organisations advocating on behalf of those communities who will be most severely impacted by climate change. This suggests that it can serve as an appropriate model to reconcile the challenge of meeting global food needs in the context of climate change. The question is whether this potential can be realised, having regard to competing economic interests.

La Via Campesina's food sovereignty model

The body of work being developed by the CSM is beginning to articulate what a food sovereignty approach to global governance would look like. Further details of the food sovereignty model are provided in the draft Declaration on the Rights of Peasants, originally proposed by the Asian membership of La Via Campesina in 2002 and currently being debated in the UN Human Rights Council. The declaration aims to secure international recognition that peasants have rights as producers of food, both for themselves and their communities (Edelman and James 2011: 93, 100). These include specific rights to the means of agricultural production as well as land, seeds and water; the right to freely determine the price of their products and the markets in which they are sold; the right to protection of traditional agricultural knowledge and traditions; and the right to resist oppression. Underpinning the declaration are principles of autonomy and self-determination for peasant communities, which are sought to be further strengthened through the assertion of a 'right to reject' certain economic forms and practices, notably land-grabbing, genetically modified (GM) seed and forms of bio-piracy (La Via Campesina 2009).

To be enforceable in the legal sense, the declaration would need to become a convention, translated into legislation at the national level and supported by well-functioning enforcement mechanisms. As is well known, proper enforcement of internationally agreed standards has long been an Achilles heel of the international human rights regime, leading to a notorious gap between word and action (Arendt 1968: 279; Douzinas 2000: 2; Evans 2005: 6).

While there are over 20 states that have legislated for the right to food and created administrative mechanisms of the sort promoted by the FAO in its Voluntary Guidelines, there are few examples of successful implementation, with strong accountability mechanisms and the achievement of substantive changes on the ground (de Schutter 2010).

Six states have incorporated the right to food sovereignty in their constitutions and some, such as Venezuela and Ecuador, have legislated for food sovereignty and created national institutions and mechanisms for the implementation of this legislation. In Venezuela, the results to date appear to have been impressive, both in terms of the participatory structures (communal councils, farmer cooperatives) for the law's implementation and the concrete outcomes achieved, notably the transition from the status of heavy food import dependency to self-sufficiency or near self-sufficiency in numerous basic food items (Schiavoni and Camacaro 2009: 133–137). In Ecuador, according to Ricardo Intriago, President of the Ecuadoran Federation of Agricultural Centres and Farmers' Organisations, and one of eight members of the Ecuadoran National Food Sovereignty Conference, food sovereignty advocates in the country are satisfied with the legislation but very unhappy with the lack of political will on the part of the government of Rafael Correa to its substantive implementation (interview with the author 15 September 2011).

Critics of the food sovereignty approach invariably charge that only large-scale and 'modern' technologies such as gene modification will enable agricultural productivity to continue to rise and thus meet global demands for food needs (Aerni 2011; Kerr 2011; Southgate 2011). In the climate change context, the adoption of farmer-driven and knowledge-intensive, region- and climate-appropriate practices of agro-ecology are being recognised as having the potential to outperform the yields of input-intensive agriculture. Less intensive methods can enhance soil fertility (and thus carbon capture) and promote climate resilience, while also creating sustainable livelihoods for rural communities – thus addressing the issue of long-term food security (Altieri 2010: 121; de Schutter 2011; Pretty 2010: 288). Agrarian reform in recent years has also been a demonstrated route to sustainable economic development in most of the world (Altieri and Toledo 2011: 604). The food sovereignty governance model *prima facie* has much to recommend it in policy terms, to meet long-term food needs in the context of a changing climate.

Conclusions and recommendations

In March 2012, the Commission on Sustainable Agriculture and Climate Change published its final report, *Achieving food security in the face of climate change*. The Commission synthesised a wide range of assessments of climate change and food security to produce their recommendations for global policymakers. Their conclusion that 'business as usual' in a global food system would not bring about food security, nor environmental sustainability, seemed surprisingly blunt (Beddington et al. 2012). Yet this authoritative document failed to mention human rights. The underlying philosophy of 'green growth' assumes, rather than challenges, the maintenance of existing configurations of political-economic power. This points to the central difficulty facing the reformed CFS. How can it reconcile the commitment of the powerful political-economic actors who have shaped the development of marketised food security and resource-intensive food systems, with the demands of food sovereignty advocates for transformative change and knowledge-intensive, climate-resilient food systems? While the evolution of the reformed CFS, as a significantly inclusive and participatory institution in the global food governance regime, will be watched with great interest, expectations of a genuine consensus on the GSF seem, for the reasons discussed above, grounded in hope rather than optimism.

Recommendations

- The work of the reformed CFS should be supported and strengthened. All governments of Member States should participate in the CFS Plenary, and intersessional activities, in good faith. This is the one global food governance institution where all stakeholders have a seat at the table, including representatives of groups who are most disadvantaged by the operation of the current global food and agricultural system. The recommendations of the HLPE in its June 2012 report on climate change and food security should be incorporated into the GFS.
- A strong commitment to a human rights approach can eradicate hunger and malnutrition and build climate-resilient food systems. There are striking examples that show that a firm commitment to the right to food and the resultant prioritisation of resources and actions can quickly reduce high levels of hunger. This requires, however, a shift away from a worldview that sees food in economistic terms as largely or solely a commodity for trade and towards a more integrated and holistic perspective that understands the social and environmental dimensions of food.
- Governments should support and encourage food-related social movements. Social movements have a critical role to play in redressing the democratic deficit of recent years, in which corporate lobbies have in many instances been able to exercise a

disproportionate influence on policy development. Food Policy Councils, food charters, food alliances and the like reveal the growth and emergence of a motivated citizenry desirous of food systems that are more fair, sustainable and climate-resilient. Governments should identify means to support these movements.

References

Aerni, P. 2011. Food sovereignty and its discontents. *ATDF journal,* 8 (1/2), 23–40.

Altieri, M.A. and Toledo, V.M. 2011. The agroecological revolution in Latin America: rescuing nature, ensuring food sovereignty and empowering peasants. *The Journal of Peasant Studies,* 38 (3), 587–612.

Altieri, M. 2010. Scaling up agroecological approaches for food sovereignty in Latin America. *In:* H. Wittman, A.A. Desmarais and N. Wiebe, eds. *Food sovereignty: reconnecting food, nature and community.* Black Point, NS: Fernwood Publishing, 120–133.

Arendt, H. 1968. *The origins of totalitarianism.* New York: Harcourt Books.

Beddington, J., Asaduzzaman, M., Clark, M., Fernandez, A., Guillou, M., Jahn, M., Erda, L., Mamo, T., Van Bo, N., Nobre, C.A., Scholes, R., Sharma, R. and Wakhungu, J. 2012. *Achieving food security in the face of climate change: final report from the Commission on Sustainable Agriculture and Climate Change. CGIAR Research Program on Climate Change, Agriculture and Food Security (CCAFS).* Copenhagen, Denmark. Available online at: www.ccafs.cgiar.org/commission, accessed 20 April 2012.

Bello, W. 2000. Why reform of the WTO is the wrong agenda. *Kasarinlan: Philippine journal of third world studies,* 15 (1), 7–22.

Bumpus, A.G. and Liverman, D.M. 2010. Carbon colonialism? Offsets, greenhouse gas reductions, and sustainable development. *In*: R. Peet, P. Robbins and M. Watts, eds. *Global political ecology.* Abingdon, Oxon: Routledge.

Cadman, T. 2011. *Quality and legitimacy of global governance: case lessons from forestry.* Basingstoke: Palgrave Macmillan.

Civil Society Mechanism (CSM). undated. *CSM working statement on the GSF.* http://www.csm4cfs.org/policy_working_groups-6/global_strategic_framework-8/gsf_draft_1-45/, accessed 8 May 2012.

CSM. 2010. *Civil Society Working Statement on the draft Global Strategic Framework for food security and nutrition,* December 2011. 6/global_strategic_framework-http://www.csm4cfs.org/policy_working_groups-8/cso_working_document_on_the_gsf-41/, accessed 8 May 2012.

Committee on World Food Security. 2012. *Global strategic framework for food security and nutrition: first draft.* FAO, http://www.fao.org/fileadmin/templates/cfs/Docs1112/WGs/GSF/CFSBriefing27Feb2012/MD067_CFS_GSF_Draft_One_En.pdf, accessed 2 May 2012.

Committee on World Food Security. 2009. *Reform of the Committee on world food security: final version.* CFS:2009/2, Rev.2.

de Schutter, O. 2011. Food crises: five priorities for the G20. *The Guardian,* 16 June 2011, http://www.srfood.org/index.php/en/component/content/article/1-latest-news/1394-g20-agriculture-5-priorities-to-end-food-crises, accessed 4 October 2011.

de Schutter, O. 2010. *Countries tackling hunger with a right to food approach: significant progress in implementing the right to food at national scale in Africa, Latin America and South Asia*. Briefing Note 1, May 2010, http://www2.ohchr.org/english/issues/food/docs/Briefing_Note_01_May_2010_EN.pdf, accessed 16 April 2012.

de Schutter, O. 2009. Governing world food security: a new role for the Committee on world food security. *Right to food quarterly,* 4 (1), 2–3, http://www.fian.org/resources/documents/categoria-3/right-to-food-quarterly-vol-4-no.1/pdf, accessed 15 August 2011.

Douzinas, C. 2000. *The end of human rights*. Oxford: Hart Publishing,

Downes, C. 2006. Must the losers of free trade go hungry? Reconciling WTO obligations and the right to food. *Virginia journal of international law,* 47, 619–692.

Edelman, M. and James, C. 2011. Peasants' rights and the UN system: quixotic struggle? Or emancipatory idea whose time has come? *Journal of peasant studies,* 38 (1), 81–108.

Evans, T. 2005. *The politics of human rights*, 2nd edition. London: Pluto Press.

GRAIN. 2010. The new farm owners: corporate investors and the control of overseas farmland. *In*: F. Magdoff and B. Tokar, eds. *Agriculture and food in crisis: conflict, resistance and renewal*. New York: Monthly Review Press, 139–153.

Harvey, D. 2003. *The new imperialism*. Oxford: Oxford University Press.

Hassan, R., Scholes, R. and Ash. N. (eds). 2005. Millennium ecosystem assessment. *Ecosystems and human well-being current state and trends*. Volume 1. Washington, DC: Island Press.

HLPE. 2012. *Food security and climate change: A report by the High-Level Panel of Experts on Food Security and Nutrition of the Committee on World Food Security*. Rome: June 2012

Holt-Giménez, E. and Shattuck, A. 2011. Food crises, food regimes and food movements: rumblings of reform or tides of transformation? *Journal of peasant studies*, 38(1), 109–144.

International Assessment of Agricultural Knowledge, Science and Technology for Development. 2008. *Executive summary of the synthesis report*. Johannesburg.

Kerr, W.A. 2011. Food sovereignty: old protectionism in somewhat recycled bottles. *ATDF Journal,* 8 (1/2), 4–9.

Klein, N. 2007. *The shock doctrine: the rise of disaster capitalism*. London: Allen Lane.

La Via Campesina. 2011. *La Via Campesina call to Durban: soil carbon is not for sale*. http://www.redd-monitor.org/2011/09/17/our-carbon-is-not-for-sale-via-campesina-rejects-redd-again/, accessed 21 September 2011.

La Via Campesina. 2009. *Draft declaration of rights of peasants – women and men*. Seoul, March 2009, http://viacampesina.net/downloads/PDF/EN-3.pdf, accessed 3 October 2009.

Lynas, M. 2009. *Six Degrees: our future on a hotter planet*. London: Fourth Estate.

Magdoff, F. and Tokar, B. 2010. Agriculture and food in crisis: an overview. *In*: F. Magdoff and B. Tokar, eds. *Agriculture and food in crisis: conflict, resistance and renewal*. New York: Monthly Review Press, 9–30.

Monbiot, G. 2004. *The age of consent: a manifesto for a new world order*. London: Harper Perennial.

Moore, J.W. 2010. The end of the road? Agricultural revolutions in capitalist world ecology, 1450–2010. *Journal of agrarian change,* 10 (3), 389–413.

Overbeek, H.W., Dingwerth, K., Pattberg, P. and Compagnon, D. 2010. Forum: global governance: decline or maturation of an academic concept? *International studies review*, 12: 696–719.

Overbeek, H. W. 2004. Global governance, class, hegemony: a historical materialist perspective. *In:* A.D. Ba and M.J. Hoffmann, eds. *Contending perspectives on global governance: coherence, contestation and world order*. London and New York: Routledge.

Peet, R., Robbins, P. and Watts, M., eds. 2010. *Global political ecology*. Abingdon, Oxon: Routledge.

Pretty, J. 2010. Can agriculture feed 9 billion people? *In*: F. Magdoff and B. Tokar, eds. *Agriculture and food in crisis: conflict, resistance and renewal*. New York: Monthly Review Press, 283–298.

Schiavoni, C. and Camacaro, W. 2009. The Venezuelan effort to build a new food and agriculture system. *Monthly review*, 61 (3), 129–141.

Sobhan, H., Guerra, A. and Rahman, A.F.M.S., eds. 2012. *The Committee on world food security: a guide for civil society*. International Food Security Network, http://www.iatp.org/files/CFS_Booklet_Final.pdf, accessed 1 May 2012.

Southgate, D. 2011. Food sovereignty: The idea's origins and dubious merits. *ATDF Journal*, 8 (1/2), 18–22.

Steinberg, Richard H. 2002. In the shadow of law or power? Consensus-based bargaining and outcomes in the GATT/WTO. *International organization*, 56(2), 339–374.

Stern, N. 2008. The economics of climate change. *American economic review: papers & proceedings, 2008*, 98(2), 1–37.

Tokar, B. 2010. Biofuels and the global food crisis. *In*: F. Magdoff and B. Tokar, eds. *Agriculture and food in crisis: conflict, resistance and renewal*. New York: Monthly Review Press, 121–138.

Weis, A.J. 2007. *The global food economy: the battle for the future of farming*. London: Zed Books.

Wilkinson, R. 2011. Measuring the WTO's performance: an alternative account. *Global policy*, 2 (1), 43–52.

Wittman, H., Desmarais, A.A. and Wiebe, N. 2010. *Food sovereignty: reconnecting food, nature and community*. Black Point, NS: Fernwood Publishing.

11
Innovation and Global to Local Energy Governance

Guilherme B.R. Lambais and Guilherme Gonçalves

Introduction

The relationship of energy to global climate change is one of the most important links to be made concerning the impact of carbon emissions on the global environment. Energy systems are a fundamental aspect of economic activity, which is currently highly dependent on fossil fuels – the combustion of which accounted for 84 per cent of GHG emissions in 2009 (OECD 2012: 11) (Figure 11.1).

There is an undeniable link between the three systems – energy, economic and earth systems – where actions and events are not confined to the domain of origin but constitute instead a complex global system (Galaz et al. 2012: 80–82). Taken separately, each can also be interpreted as complex systems themselves, meaning that between and within systems the dynamics of interaction are non-linear and also multi-scale in both space and time.

The earth system is a major domain, with several natural subsystems each linked to economic and energy systems, but in ways that are not yet completely understood (Andrade and Romeiro 2011: 6–9). Arrow et al. (1995: 520) have for long argued that there are limits to the planet's carrying capacity, that is, its ability to generate material production in the future. In addition, 'Carrying capacities in nature are not fixed, static, or simple relations. They are contingent on technology, preferences, and the structure of production and consumption. They are also contingent on the ever-changing state of interactions between the physical and biotic environment' (ibid: 521). Accordingly, human activity can influence environmental dynamics in unpredictable ways. This has led to what has been referred to as the 'planetary boundaries' framework (see also Chapter 9, this volume).

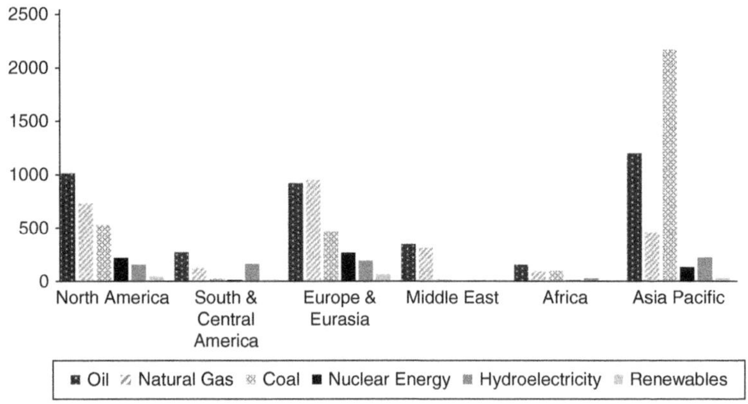

Figure 11.1 World consumption of energy in 2009
Source: Prepared by the authors with data from the BP Statistical Review of World Energy (2011)
Notes: Oil consumption is measured in million tonnes; other fuels in million tonnes of oil equivalent.

These boundaries provide a reference point for safe human operations within the planet's biophysical subsystems or processes (Rockström et al. 2009: 472).

The energy and economic systems likewise interact in a dynamic fashion; the economy is a domain with subsectors where energy is both an input for all sectors and a sector itself. Well-established technologies dominate the energy system due to positive feedbacks with economic systems and (due to increasing returns and high capital and physical investments) within the energy system itself – all of which results in a 'carbon lock-in' between both systems (Arrow 2003: 32; Unruh 2000: 817–830; Varnäs et al. 2012: 5–6). These technologies have been attached to the energy regime complex (Colgan et al. 2011). When examined through an institutional and history-based lens, this complex leads to the pessimistic normative conclusion that neither the structures of participation nor the processes of deliberation are particularly legitimate.

Coupling technology and the environment, Arrow et al. (1995: 521) argue that on account of human innovation and biological evolution an exact figure for the limits of the earth's carrying capacity is unknown. Therefore, it is suggested here that the promotion of innovation in renewable energies represents a significant contribution towards solving the conundrum of maintaining relative economic growth, while inducing positive changes within the current energy system. In order to achieve better governance at a global scale, Walker et al. call for 'a renewed focus on effective cooperation, facilitated by better-designed institutions', in contrast to existing 'international institutions [that] primarily focus on single problems, ignoring system-wide interactions' (2009: 1345–1346). Responding to this view, Rao (2009) argues that instead of creating new formal global institutions, it would be 'more useful to examine the problem in terms of bottom-up approaches', leading to a better analysis of the effects of incentives, like differential taxation and subsidies, on producers and consumers (that is, supply and demand). Furthermore, Cash et al. (2003: 8086–8091) assert that science and technology must play a more central role in sustainable development, and the study suggests that 'efforts to mobilize science and technology for sustainability are more likely to be effective when they manage boundaries between knowledge and action in ways that simultaneously enhance the salience, credibility, and legitimacy of the information they produce' (ibid: 8086).

In view of these observations, this chapter critiques existing energy governance and argues for an alternative model that emphasises multi-level, multi-stakeholder participation in energy governance systems. These are more suited to enabling the co-evolution of supply and demand within processes of energy generation and consumption at the local level, thus transforming prices and preferences and thereby stimulating renewable energy markets. New 'smart grid' technology is particularly relevant to local systems and can be regionally organised, making a valuable contribution towards renewable energy solutions.

In the following section, the energy regime complex is explored. A historico-institutional analysis follows, which argues that global governance of energy is not likely to deliver legitimate change towards renewable energy. This is followed by a presentation of new options for energy governance based on a bottom-up approach and some final comments are made by way of conclusion.

The regime complexes for energy and climate change

Following Keohane and Victor (2010), there is agreement that the institutions that govern energy also qualify as a regime complex. The energy regime complex, however, predates that of climate change and only partially overlaps, although both resemble each other due to fragmentation of interests and power, leading to a pattern of punctuated equilibrium, that is, characterised by periods of significant innovation and of stasis, as opposed to a continuous, gradual process of change (Colgan et al.

2011: 118–199). Change in this regime complex is also, like climate change, likely to be path-dependent, driven by satisfaction or dissatisfaction related to oil prices and resulting homogeneity or heterogeneity among actors. Path-dependent change occurs in existing institutions when interests of dissatisfied actors are homogeneous. If interests of these dissatisfied actors are heterogeneous, incremental change does not command sufficient support; states seeking major changes have to create new institutions. The functions of these new institutions, however, are not likely to deviate much from immediate previous institutions (Colgan et al. 2011: 119–122). The processes of innovation and stasis with these organisations are reproduced in Figure 11.2 (Colgan et al. 2011: 123–130):

- **1950–1965**: During this period, there was almost no structured cooperation among major actors. The only multilateral institution was the International Atomic Energy Agency (IAEA), the establishment of which grew out of a fear of an arms race and the enthusiasm for a new source of power. The lack of international institutions indicates that until the recovery from the Second World War almost all energy consumed was produced within borders. With fast economic growth, however, Europe shifted from a coal-based economy to one based on imports of oil from the Middle East. The principal oil consumers created the Organisation for Economic Cooperation and Development (OECD), and this initiative was accompanied by the creation of the Organisation of Petroleum Exporting Countries (OPEC).
- **1965–1980**: OPEC was initially created to emulate pro-ration policies from the United States, which in 1970 was by far the largest oil producer in the world and 85 per cent self-sufficient. Initially OPEC was not a cartel, but a way to reduce its members' dependence on international oil companies by developing policies regarding royalties and taxes. It should be noted that a variant, OAPEC, was created in 1968 by Kuwait, Saudi Arabia and Libya, paving the way for the 1973 oil boycott and production cuts. By 1973, however, OPEC had 12 members and was producing 54 per cent of world oil output, definitively shifting the distribution of power in the international arena in favour of OPEC members and their state-led oil companies. In response, the International Energy Agency (IEA) was created in 1974 with 16 members as an autonomous agency by the predominantly Western powers-dominated OECD. IEA was created with two main functions: (a) maintaining and improving the system for coping with oil supply disruptions; (b) acting as a body for development

of policy, information sharing and technology transfer. It should be noted that in the previous year, the Latin American Energy Organisation (OLADE) was also created, a productive initiative, but it has been modest in its accomplishments to date. Alarmed by the 1970s energy and monetary crises, the Group of Seven (G7) was also formed in this period, although it has concentrated primarily on macroeconomic issues, focusing mainly on tracking oil price fluctuations.

- **1980–1995**: During this period, IEA became OECD member-wide, but evolved into a rigid and inequitable body, offering little incentive for change. In addition, consumer countries increasingly attempted to shift away from OPEC's oil, due to ongoing supply crises, which led to the creation of two new multilateral energy initiatives in the early 1990s. First, the International Energy Forum (IEF), a biannual inclusive dialogue between energy consumers and producers, amounting to 80 countries, including all IEA and OPEC members, giving voice to countries outside the traditional Western ambit, namely Brazil, Russia, Mexico, India, China and South Africa. Its main achievement to date has been to increase the transparency and availability of oil-supply data. The other new institution created was the Energy Charter Treaty (ECT), but this is not especially preeminent, as it includes only Eurasian countries.
- **1995–2010**: Following an increasing demand for other forms of energy, the International Renewable Energy Agency (IRENA) entered into force in 2010. Its main function is one of 'soft coordination', gathering and disseminating knowledge, providing policy advice and promoting technology transfer and research. The other institutional newcomer is the International Partnership on Energy Efficiency Cooperation (IPEEC). Arising from the Group of Eight, including Russia (G8) with support from China, India, South Korea, Brazil and Mexico, it is limited to the exchange of information on best practices and joint research and development. Meanwhile, the IEA has shifted its attention to climate change, providing the Secretariat for the Annex I country signatories of the United Nations Framework Convention on Climate Change (UNFCCC).

The partial conclusions that can be drawn from history are that the energy regime complex is somewhat ossified, dominated by few and resistant to change, resulting in patterns of punctuated equilibrium. A more in-depth institutional analysis of the global governance of energy, using the indicators of Cadman (2011), follows below.

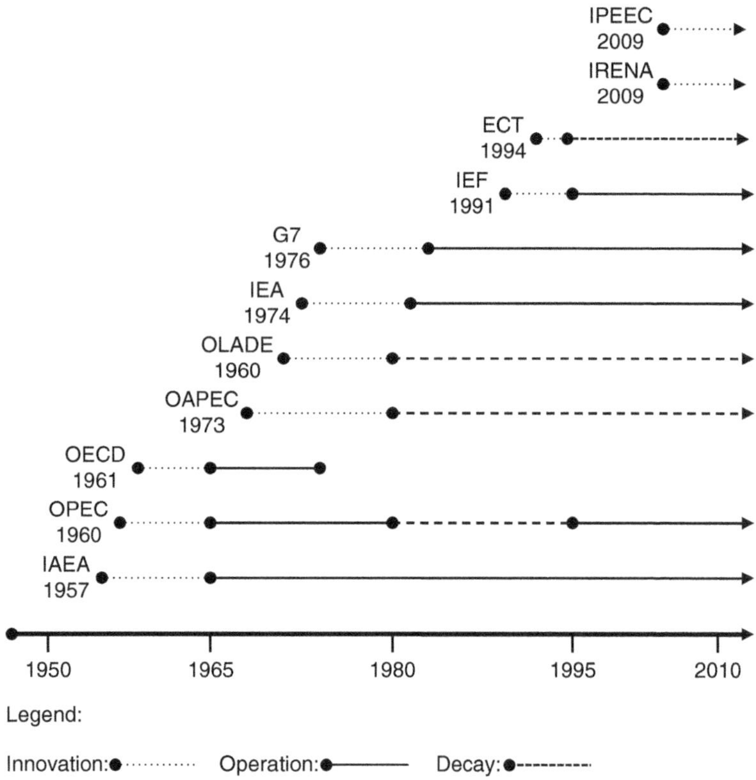

Figure 11.2 Institutional changes in the energy regime complex
Source: Colgan et al. (2011: 130)

Institutional analysis of current energy governance

Governance has been defined as 'control exercised by a variety of public and private institutions that have been established at different spatial scales' (Perrons 2004: 255). As argued immediately above, the governance of the energy system is directly related to formal organisations at the global level and shows remarkable stability over time, concentrating power and resources in the hands of few. In turn, the institutions that have emerged or have been consciously designed also respond to very few interests.

Florini and Sovacool summarise the energy history from a governance point of view. They argue that governments and other interests have not been effective in trans-boundary organisation, and service provision exists in a distorted and poorly regulated market, which, although technically domestic, is effectively international. However, the system's ability to address the externalities associated with different energy commodities and move to a low-carbon market is weak. Its governance is also compromised by conflicting national policies and lack of investment, which can lead to price volatility. National governments also intervene in energy markets in order to gain energy security and independence, yet fail to take action on climate change and other important aspects of trans-boundary atmospheric pollution (such as acid rain or emissions of mercury). Governance failings are therefore both national and international in scope. As a consequence, 'few if any governments are well structured to govern energy issues, much less participate in effective systems of global energy governance' (2009: 5239).

An analysis across the historical development of the regime complex permits a number of observations about energy governance. First, in relation to the principle of meaningful participation:

- Inclusiveness: IEF has begun to include more members beyond the initial 80 members, but IEA and OPEC members still dominate. Real inclusiveness only occurred relatively recently, with IRENA entering into force (it consists of 149 countries plus the EU but is ratified by only 65 countries).
- Equality: If equality is understood as an equal division between economic, social and environmental interests, the regime complex is visibly unequal: economic interests largely determine the agenda.
- Resources: Considering that US$557 billion are globally spent to subsidise fossil fuels (Fairley 2011: 2), it is clear that resources are skewed in their favour, at the expense of renewable energy.
- Accountability: If any accountability is present at all, it generally only exists within state-based clubs. OPEC members often hold the rest of the world hostage and act in an opportunist manner to bypass oil supply regulations, usually successfully. Or else the Middle East is subjected to interference from other global superpowers (mostly the United States), also with no accountability.
- Transparency: The inner workings of energy production deal making are very opaque. The only measures towards increasing transparency are to be found in the increase in the quality and availability of oil data provided by the IEF.

Finally, in relation to the principle of productive deliberation:

- Democracy: The ways in which the energy regime complex acts cannot be classed as democratic. It is better classified as a plutocracy, in which the actors, who control the resources, determine the outcomes.
- Agreement: The effectiveness of agreements varies according to their ambit: the 'softer' the issues such as promoting coordination, or disseminating information and policy advice, the greater the level of agreement; the 'harder' the issues, such as setting quotas for subsidies or oil supplies, the smaller the chance they have of being universally supported.
- Dispute settlement: At the point of writing, it was not possible to determine what mechanisms were in place for some of the newer bodies. In the case of pre-2000 institutions, historical examination reveals that these institutions have generally been quite inflexible and shown little favour to parties with whom they are in dispute.
- Behavioural change: It is difficult to determine behavioural change. What can be argued is that the regime complex functions in a pattern of punctuated equilibrium with some times of inertia and others of relatively rapid change. However, due to high path-dependence a commitment to making the necessarily rapid transition to greater adoption of renewable energies is not apparent at this time.
- Problem solving: In terms of reducing fossil fuel emissions associated with human-induced climate change, efforts have been isolated and driven by action at the 'soft' level; problem-solving activities have been more related to ensuring the smooth functioning of fossil fuel markets than replacing these markets.
- Durability: If recent history is a guide to the durability of the global energy regime complex, it might be said that an end to fossil fuel dependence is nowhere in sight. The status quo is proving to be extremely durable.

Building on the analysis of Florini and Sovacool (2009), it is clear that the energy regime complex as it is will not produce multi-stakeholder governance with inclusiveness of actors, equality of interests and distribution of resources; nor is there much prospect of greater accountability, although there have been some small improvements at the peripheries of transparency. Nor is there likely to be a democratic system with effective agreement and dispute settlement any time soon. Behavioural change towards a greater dependency on renewable energy, thus contributing to the reduction of GHG emissions, still seems a long way off.

Future options for energy governance – the importance of innovation

By way of contrast to the existing complex, some new options for energy governance, built upon evolutionary system dynamics, bottom-up approaches to economics and the promotion of technological innovation are explored here. Creating a viable multi-level governance system based on innovation and co-evolution of supply and demand must not be a distant future anymore (Kemp et al. 2007: 78–80). Some of the dynamics of global-scale incentives count in favour of alternative energy sources, and there are opportunities for bottom-up changes to complement longer-term structural changes at the global level (Acemoglu et al. 2012: 131–134). Bardhan has suggested that 'technological changes have also made it somewhat easier than before to provide public services (like electricity and water supply) relatively efficiently in smaller market areas, and the lower levels of government have now a greater ability to handle certain tasks' (2002: 187). Regional multi-stakeholder systems could be developed within an institutional setting of norms, routines and conventions concerning the organisational support infrastructure for economic development. This would minimise transaction costs and take advantage of geographical proximity to generate economies of scale. This would in turn build trust and facilitate institutional, organisational and individual innovation and information exchange (Cooke et al. 1998: 1563).

Such systems might take a number of forms. On the supply side, *sectoral systems* of innovation would consist of institutional interactions at various levels of aggregation, through processes of communication, exchange, cooperation, competition and command (Malerba 2002: 247). Alternatively, a more *socio-technical* approach places a greater emphasis on the demand side of energy innovation, embracing the linkages between societal needs for transport, communication, nutrition and so forth, with the technical means to achieve these social goals (Geels 2004: 900–904). This approach also has the advantage of incorporating aspects of innovation beyond the economic sphere, such as individual and social learning for sustainable development (Siebenhüner 2000: 15: 25). Adoption and assimilation of technology by means of 'learning by doing' and 'learning from others' create favourable conditions for innovation (Conley and Udry 2010: 35; Foster and Rosenzweig 1995: 1176–1179; Vieira Filho and Silveira 2012: 267–268). Communities can also experiment with various forms of ownership, promoting cooperative arrangements, which are known to be more equitable (Pollin 2012: 12–14) and which could overcome initial problems of financing innovation (Perez 2004: 775–800). If the right incentives are put in place, there could be an advancement of social preferences towards the evolution of public goods (Bowles and Hwang 2008: 1811–1813).

In terms of existing technologies, smart grids, which already exist off-the-shelf, constitute an ideal core infrastructure for regional energy networks at the urban level, as argued by the IEA (2009: 15–17) and, moreover, by the 2011 report about smart grids. Smart grids allow for the incorporation of different and variable energy sources into electricity networks by means of digital and other technologies. They coordinate and integrate energy generators, operators and end users and other market stakeholders within a system optimised for energy efficiency, thereby reducing costs and impacts on the environment while increasing reliability, resilience and stability at the same time. The very nature of such systems encourages solution-oriented cooperation among all stakeholders along the supply chain from manufacturers of equipment to end consumers (IEA 2011: 6–7). Smart-grid technology constitutes a material base to human systems of governance, which embody the kind of democratic, collaborative interaction suggested by Cadman (2011: 8).

Current systems of energy governance have resulted in varying degrees of antagonism between economic, social and political interests, and a more equitable integration of the demands of different social and economic actors and interest groups is required. Such integration is most feasible within a multi-level and multi-stakeholder setting. But such a setting requires decisions to be made across horizontal scales (space) and vertical scales (levels of organisation). This in turn generates cross-scale institutional linkages, comprised of a diversity of agents (for example firms and governments) (Berkes 2002: 293). The interests of specific economic and political groups must not supplant information, empirical evidence, scientific and technical knowledge derived from researchers and science and technology institutions. To enhance salience across institutional contexts and ensure legitimacy, local and regional communities need to have representation as well, as this also allows for a sufficiently wide variety of visions that comprise science.

Such a collective construction of energy governance could be based around some form of non-partisan and non-profit representative body. This facilitating organisation should encourage debate and participation by stakeholders, through various means of dialogue channels, discussion forums and public hearings. Thus, this structure should also define the deliberative processes that set the economic parameters of governance, such as taxation, levels and beneficiaries of subsidies, public research and development financing, and models of intellectual property and property ownership. Hall and Helmers (2010) discuss the effectiveness of intellectual property models for clean technology transfer, concluding that given the need for local adaptation, patent protection may not be the optimal instrument for encouraging innovation. They also confirm (2011) that 'hybrid' forms of knowledge management, which combine both public innovation and private patenting, encourage the development and diffusion of climate change-related technologies.

Given these considerations, the establishment and evolution of the supply and demand of clean energy through regional and community-based smart grids should be given serious consideration as an alternative model of energy governance. Supply of energy by agents (that is, individuals or organisations) could then be made through existing technologies, like solar photo-voltaics (Varnäs et al. 2012) or through yet-to-be invented technologies, since the socio-economic environment will be favourably set to foster innovation. Likewise, demand for this energy would be guaranteed by regional and local markets, instead of the depending on the uncertainty and instability of national and international markets.

Conclusions and recommendations

Energy systems are intrinsically linked to both economic systems and earth systems. Given these inter-linkages, such systems are highly complex and changes in one or other system can have unpredictable global consequences, both in terms of feedback, amplification and/or reverberation affecting all systems. Policies dealing with such complexities require alternative approaches to energy management in the context of climate change. An investigation of the history of the current energy regime complex shows that the existing structures and processes governance are not likely to move towards the necessary uptake of renewable energy required for responding to human-induced climate change.

Alternatively, bottom-up initiatives could be developed using technological innovation and evolutionary system dynamics. Regional and local renewable energy markets based on smart grids have the potential to create favourable socio-economic environments for innovation, co-evolution of supply and demand and increased economic returns associated with the diffusion, adoption and assimilation of new technologies. These markets would evolve from a form of local governance that is more legitimate than current national and international energy markets, which are known to concentrate power and result in instabilities.

The governance agenda set forth in this chapter would both control and reduce energy pollution. Phasing out the current dangerous forms of energy generation and their associated environmental and health risks would encourage the move towards eco-efficient energy generation and ultimately dismantle the current paradigm. This requires broad-scale social and political commitment to innovations in the processes, equipment and inputs associated with the generation and distribution of clean and cheap energy. This will also lead to a transformation in economic systems, but that is a subject for future discussions.

References

Acemoglu, D., Aghion, P., Bursztyn, L. and Hemous, D. 2012. the environment and directed technical change. *American economic review*, 102(1), 131–166.

Andrade, D.C. and Romeiro, A.R. 2011. Degradação ambiental e teoria econômica: algumas reflexões sobre uma 'economia dos ecossistemas'. *EconomiA, Brasília (DF)*, 12 (1), 3–26.

Arrow, K. 2003. Path dependence and competitive equilibrium. *In:* W. Sundstrom, T. Guinnane and W. Whatley, eds. *History matters: essays on economic growth, technology, and demographic change*. Stanford, CA: Stanford University Press, 23–33.

Arrow, K., Bolin, B., Costanza, R., Dasgupta, P., Folke, C., Holling, C. S., Jansson, B.-O., Levin, S., Mäler, K.-G., Perrings, C. and Pimentel, D. 1995. Economic growth, carrying capacity and the environment. *Science*, 268, 520–521.

Bardhan, P. 2002. Decentralization of governance and development. *The journal of economic perspectives*, 16 (4), 185–205.

Berkes, F. 2002. Cross-scale institutional linkages: perspectives from the bottom-up. *In:* E. Ostrom, T. Dietz, N. Dolsak, P. Stern, S. Stonich, and E. Weber, eds. *The drama of the commons*. Washington, DC: National Academy Press, 293–322.

Bowles, S. and Huang, S-H. 2008. Social preferences and public economics: Mechanism design when social preferences depend on incentives. *Journal of public economics*, 98 (8–9), 1811–1820.

BP Statistical Review of World Energy. 2011. http://www.bp.com/liveassets/bp_internet/globalbp/globalbp_uk_english/reports_and_publications/statistical_energy_review_2011/STAGING/local_assets/spreadsheets/statistical_review_of_world_energy_full_report_2012.xlsx, accessed 31 May 2012.

Cadman, T. 2011. *Quality and legitimacy of global governance: case lessons from forestry*. Basingstoke: Palgrave Macmillan.

Cash, D.W., Clark, W.C., Alcock, F., Dickson, N.M., Eckley, N., Guston, D.H., Jäger, J. and Mitchel, R.B. 2003. Knowledge systems for sustainable development. *Proceedings of the national academy of sciences of the United States of America*, 100 (14), 8086–8091.

Colgan, J.D., Keohane, R.O. and Van de Graaf, T. 2012. Punctuated equilibrium in the energy regime complex. *The review of international organizations*, 7 (2), 117–143.

Conley, T.G. and Udry, C. 2010. Learning about a new technology: pineapple in Ghana. *American economic review*, 100 (1), 35–69.
Cooke, P., Uranga, M.G. and Etxebarria, G. 1998. Regional systems of innovation: an evolutionary perspective. *Environment and Planning A*, 30 (9), 1563–1584.
Fairley, P. 2011. Next generation biofuels. *Nature*, 474 (7352), S2–S5.
Florini, A. and Sovacool, B.K. 2009. Who governs energy? The challenges facing global energy governance. *Energy policy*, 37 (2), 5239–5248.
Foster, A.D. and Rosenzweig, M.R. 1995. Learning by doing and learning from others: human capital and technical change in agriculture. *Journal of political economy*, 103 (6), 1176–1209.
Galaz, V., Biermann, F., Crona, B., Loorbach, D., Folke, C., Olsson, P., Nilsson, M., Allouche, J., Persson, A. and Reischi, G. 2012. 'Planetary boundaries' – exploring the challenges for global environmental governance. *Current opinion in environmental sustainability*, 4 (1), 74–79.
Geels, F. W. 2004. From sectoral systems of innovation to socio-technical systems Insights about dynamics and change from sociology and institutional theory. *Research policy*, 33 (6–7), 897–920.
Hall, B.H. and Helmers, C. 2011. Innovation and diffusion of clean/green technology: can patent commons help? NBER working paper 16920, 1–33.
Hall, B.H. and Helmers, C. 2010. The role of patent protection in (clean/green) technology transfer. NBER working paper 16323, 1–36.
IEA. 2011. *Technology roadmap: smart grids*. Paris: OECD/IEA, 1–48.
IEA. 2009. *Cities, towns & renewable energy*. Paris: OECD/IEA, 1–186.
Kemp, R., Derk, L. and Rotmans, J. 2007. Transition management as a model for managing processes of co-evolution towards sustainable development. *International journal of sustainable development & world ecology*, 14 (1), 78–91.
Keohane, R.O. and Victor, D.G. 2010. The regime complex for climate change. *The Harvard project on international climate agreements*. John F. Kennedy School of Government (discussion paper 10–33).
Malerba, F. 2002. Sectoral systems of innovation and production. *Research policy*, 31 (2), 247–264.
OECD. 2012. *Energy*. OECD Green Growth Studies, OECD Publishing, http://dx.doi.org/10.1787/9789264115118-en, accessed 31 July 2012.
Perez, C. 2004. Finance and technical change: a neo-schumpeterian perspective. *In:* H. Hanusch and A. Pyka, eds. *The Elgar companion to neo-schumpeterian economics*. Cheltenham: Edward Elgar, 775–800.
Perrons, D. 2004. *Globalization and social change: people and places in a divided world*. London: Routledge.
Pollin, R. 2012. Public policy, community ownership and clean energy. *Cambridge journal of regions, economy and society*, 1: 18, doi: 10.1093/cjres/rss007, accessed 31 July 2012.
Rao, P.K. 2009. Missing formal institutions? Response to Walker et al. (2009), http://www.sciencemag.org/content/325/5946/1345.summary/reply#sci_el_126, accessed 9 July 2012.
Rockström, J., Steffen, W., Noone, K., Persson, A., Chapin, F.S., Lambin, E.R., Lenton, T.M., Scheffer, M., Folke, C., Shellnhuber, H.J., Nykvist, B., Wit, C.A., Hughes, T., Van der Leeuw, S., Rodhe, H., Sörlin, S., Snyder, P.K., Costanza, R., Svedin, U., Falkenmark, M., Karlberg, L., Corell, R.W., Fabry, V.J., Hansen, J., Walker, B., Liverman, D., Richardson, K., Crutzen, P. and Foley, J. 2009. A safe operating space for humanity. *Nature*, 461 (7263), 472–475.

Siebenhüner, B. 2000. From homo economicus to homo sustinens – Towards a new conception of man for ecological economics. *Ecological economics*, 31 (1), 15–25.

Unruh, G.C. 2000. Understanding carbon lock-in. *Energy policy*, 28 (12), 817–830.

Varnäs, A., Fahnestock, J., Nykvist, B., Chandler, C., Erickson, P., Nilsson, M., Han, G., Lazarus, M. and Hallding, K. 2012. *Driving technological innovation for a low-carbon society*. Research Report, Stockholm Environment Institute.

Vieira Filho, J.E.R. and Silveira, J.M.F.J. da. 2011. An evolutionary model of learning in agriculture. *Brazilian journal of innovation*, Campinas (SP), 10 (2), 265–300.

Walker, B., Barrett, S., Polasky, S., Galaz, V., Folke, C., Engström, G., Ackerman, F., Arrow, K., Carpenter, S., Chopra, K., Daily, G., Ehrlich, P., Hughes, T., Kautsky, N., Levin, S., Mäler, K-G., Shogren, J., Vincent, J., Xepapadeas, T. and Zeeuw, A. 2009. Looming global-scale failures and missing institutions. *Science*, 325 (5946), 1345–1346.

12

Climate Change, Population Movements and Governance: Case Studies in Response Mechanisms

Richard Hil

Introduction

The impacts and implications of anthropogenic climate change on ecosystems, bio-diversity and human populations are shaping political and policy discourses around the world. The challenges of mitigation and adaptation have generated urgent public debates about the most effective, secure and socio-culturally sensitive ways of reducing the threat of human-induced climate change and responding to its immediate and anticipated consequences. One of the gravest threats to humankind lies in the impact of climate change on human settlements and the potential for radically altered climatic conditions to lead to large-scale population movements, border disputes, resource wars and spatial/territorial conflicts within and between regions and nation states (Evans 2010: 5–7). Equally important are issues related to national sovereignty, identity and self-determination, all of which are being considered in a fluid and unpredictable context because of the uncertainties associated with the precise nature, scale and pace of altered climatic conditions.

The scale of the problem is immense with up to 10 per cent of the world's population at extreme risk of life-altering climate change, rendering homelands vulnerable to rising sea levels and desertification, among many other climatic threats (Global Humanitarian Forum 2010: 3). Against this background, analytical frameworks for assessing the efficacy of the quality and legitimacy of policy mechanisms are a useful tool for evaluating 'new governance' in a globalised context – especially the 'co' arrangements for participation within an institution's structure and deliberation via its processes (Cadman 2011: 1–5). This 'non-statist' and deliberative understanding of governance raises important questions about how specific governance mechanisms can be employed effectively to meet the challenges of climate-induced migration and associated disputes. Such an approach also enables an analysis of how power is exercised through institutional practices framed by geo-political relations. Arguably, without an appreciation of the latter, there is the danger of reproducing neo-colonial practices that further marginalise and exclude already disempowered populations. An example of this is the use of social and economic rather than military or other policies to assert the interests of powerful states over other nations (Nkrumah 1967). At the very least, the interests and accounts of displaced and risk-prone populations should be considered in terms of how socially just governance mechanisms are framed in the context of a reconfigured world. This becomes most apparent when the policy and process mechanisms for facilitating population movements are considered, although they extend to many other areas of concern.

The task of ensuring that governance mechanisms are applied equitably and rigorously to the development of specific policies and practices in respect of displacees rests on the means undertaken to ensure interest representation, inclusiveness, equality and resources, accountability and transparency, and fully consultative decision-making and implementation (Cadman 2011: 1–5). Each of these areas demand attention to structures that ensure the participation of key stakeholders in any deliberative process that minimally should ensure equity, access, inclusion and participation as indicators of good governance. In the context of this chapter, commentary on such matters has proceeded with several considerations in mind, particularly the inclusion of those democratically representative members of groups impacted most directly by climate change. Several questions arise in this regard, including the devolution of responsibility for decision-making in accordance with the accepted cultural traditions and mores of the groups concerned. This is an important point given that prescriptions for quality governance can themselves include suggested arrangements that do not accord with the orientations of displaced populations. In effect, this means that prior to any attempt to apply governance frameworks discussion is required in relation to the particular cultural orientations of various stakeholders. In so doing, attention will be drawn to cultural differences and possible avenues for agreed policy and procedural methods that are cognisant of cultural diversity. Thus, the 'weight' or 'the extent to which influence is (equally) distributed among the active participants' must remain attuned to cultural differences and diverse views in respect of ameliorative action (Cadman 2012: 16).

The imposition of governance frameworks without such cultural considerations – and, by inference, the power held by various stakeholders in terms of their capacity to influence decisions – will inevitably result in imposed 'solutions' that may lead to all manners of unintended consequences. This chapter examines issues relating to the role played by key stakeholders in terms of the representation of the interests of 'ordinary people' in deliberative processes. It offers some critical insights into the multiple and competing challenges facing organisations and institutions as they seek to respond to altered geo-political relations occasioned by population movements and allied questions of cross-border security and national sovereignty. Attention is restricted largely to questions of principle, and particularly the extent to which those impacted by climate change participate meaningfully in decisions that shape their lives. It focuses on two case studies of countries that are, or will be, significantly impacted by climate change – Papua New Guinea and Italy – and discusses the implications for those populations most affected. While only one of the case studies outlined in this chapter (Papua New Guinea), with its heavy reliance on secondary sources, deals tangentially with the matters raised above, it nonetheless indicates that to some degree interest representation among those impacted by climate change did occur even though the precise nature of selection was not apparent.

It is further argued that the existential insecurities instigated by climate change necessarily render problematic traditional understandings of sovereignty, the nation state and cross-border relations. If those engaged in policymaking are to produce some positive benefits, they must remain mindful of the globalised nature of climate change and the historicised nature of geo-political and power relations. This means that all parties remain alert to and cognisant of the origins of the current crises facing populations impacted by climate change as well as the processes put in place to facilitate ameliorative 'solutions'.

Conflict, future–present

Despite critiques from various quarters, including doubts over initial motivation, methodological techniques and conclusions drawn from the available evidence (Dunlap and McCright 2011: 240–56), climate science is now at a high degree of certainty in respect of predicted climactic conditions, and there is general consensus there will be an overall rise in the earth's surface temperature, sea level rises, desertification, more intense rainfall, storms and other extreme events (Hansen 2010: ix–xiv). The environmental results of such changes are profound in terms of the impact on bio-diversity, the loss of 'carbon sink' forests, melting of polar icecaps and permafrost, with the consequent increase in carbon emissions (IPCC 2010: 5–8 CSIRO and BOM 2012: 3–12). The secondary social, economic, cultural and political effects are incalculable with the threat of food and water shortages at the forefront of current global concerns (see Chapters 9 and 10). Given the uncertainty associated with climate change, it is difficult to predict its specific effects on countries let alone plan for effective ameliorative measures (Gresser 2011: np). James Lovelock (2009), Jim McKibben (2010), Tim Flannery (2010) and Clive Hamilton (2011) and others have taken what might be termed a radical pessimistic outlook in respect of climate change, asserting that the planet is fast approaching or has already passed the point of no return. The result, it is claimed, will be devastation to and degradation of impacted lands, flooding and inundation, the spread of diseases and tens of millions of deaths as well as massive population movements, cross-border tensions and resource wars.

There are already signs in Africa, parts of Asia and the South Pacific of significant movements of people from one region to another as a result of climate change (IOM 2009: 245–301; ADB 2012: viii–ix). As recently as January 2011, the European Union (EU) debated the implications of population movements occasioned by climate change in North Africa and Southern Europe, and the United States has engaged in high-level discussions about the likely security implications of climate-induced migration (Powel 2012: np). Additionally, climate change has already caused countless deaths, territorial disputes (as in Greenland and the Northwest Passage with Canada and Russia), floods (in many parts of Australasia) and the search by low-lying island nations for relocation of populations, as in the case of Tuvalu, Kiribati and the Maldives Islands (Park 2011: 1).

National borders become increasingly irrelevant in what David Spratt and Phillip Sutton have referred to as climate catastrophe, since rising temperatures and sea levels impact all nations (2008: 3–7). However, these impacts play out most destructively on some of the world's poorest nations, which have contributed little themselves to anthropogenic climate change. Low-lying counties, atoll Pacific Islands and swathes of already arid regions in sub-Saharan Africa and Asia face the brunt of climate change. Bangladesh, India, China, Honduras, Nicaragua, Haiti, the Dominican Republic, Vietnam and the Philippines are among the countries most affected by climate change and, because of structural poverty and disadvantage, are the least able to provide the necessary support and adaptive measures required to cope with such developments (Friedman 2011: np). Population movements as a result of climate change have already begun from places like North Africa, sub-Saharan Africa and Papua New Guinea although, as noted below, the process of relocation presents many challenges to governing bodies.

In Chapter 13, Andrea Berringer highlights some of the major challenges faced by climate displacees. The root word, displace, suggests that climate displacees are compelled to move from their home/land because of reasons beyond their control, thus rendering such actions involuntary and often at the behest of other parties who may or may not take responsibility for either the causes of movement and/or the subsequent attempts at resettlement. Even where responsibility is accepted, the question arises as to the role that various stakeholders play in such decisions. Further, there is a very real human side to resettlement that relates to questions of human security, psychological well-being and cultural identity. The act of movement is thus constituted as only one transformative aspect of a complex process of dislocation/relocation that involves anticipatory and reactive grieving and loss, impacts of socio-cultural identify formation, socio-economic hardship, disruption to key institutions, civil society and social order (Doherty and Clayton 2011: np).

Climate change, altered landscapes and geo-political conflict

Predictions of population movements resulting from climate change are many and varied. At best, estimates of such future movements resulting from climate change are guesswork (DRCMGP 2008: 1). The variations in these predictions result largely from differing assumptions, methodologies and competing discursive categories. Additionally, the universal assumption that populations will inevitably move as a result of changing climatic conditions is unfounded since the lack of somewhere to go, obstacles to long distance migration, economic and political constraints and the desire to remain in one's country of origin are powerful countervailing influences (ibid: 3–4). Such analytical problems are further compounded by the uncertainties associated with future climate change developments and their effects on human populations. As such, it remains largely unclear as to when and how many displacees will be forced to move within or beyond the borders of the nation state.

Some reports suggest that the numbers of climate displacees are rising exponentially and, taken together with other environmental migrations, aggregations are said to now exceed those displaced by war (UN University 2011: np).

Drawing on a variety of information sources, the Environmental Justice Foundation (EJF) estimated the numbers of those facing extreme risk because of climate change, and therefore vulnerable to involuntary geographical movement, as somewhere between 500 and 600 million people or 10 per cent of the world's population (2008: 4). Recently the Asian Development Bank (ADB) observed that over the past two years alone, 42 million people in the Asia-Pacific region have been displaced as a result of severe weather events, many attributed to climate change (2012: viii). Desertification and environmental degradation have over the past two decades or so led at least 10 million people in Africa to move from their homelands in search of food, water and arable land (EJF 2008: 4–5). Other countries such as Bangladesh, Papua New Guinea and low-lying islands in the South Pacific such as Tuvalu, the Solomon Islands and Kiribati have also experienced population movements although, as noted below, the example of Papua New Guinea (PNG) demonstrates the many problems besetting resettlement of climate displacees. Given the threefold increase in extreme weather events over the past 30 years the IOM has noted an escalating trend in forced population movements, especially in poor countries in the global South. Twenty million people were displaced in 2008 by extreme weather events, and droughts have been experienced by over a billion people, thus forcing many, for reasons of water and food insecurity, to flee to other areas (IOM 2011: np).

Inevitably the growing incidence of population movements has generated heated public debate about the likelihood of future conflicts. Conflicts over food and water as well as various social and economic resources are seen as inevitable in contexts where host governments have to 'balance' the interests of domestic populations with those of new arrivals (Raleigh et al. 2009: 1–2; Tsuma 2011: 3–6). In poorer nations, conflicts may arise because governments cannot afford to provide the required infrastructure and support services. In cases where richer nations may absorb displaced populations questions of resource allocation and distribution are accompanied by numerous security and law and order concerns. As Elliot points out, such concerns are prevalent in most host nations, including those of richer countries. A range of causes for conflict has been identified in government and NGO reports. These include competition over scarce resources; lack of jobs or other means of economic support; rising burdens on societal infrastructure;

and cultural differences arising from nationality or ethnicity. These are all more likely to occur in locations where social instability already exists, and where there is low adaptive capacity (Elliot 2011: 7).

Elliot further argues that much of the discourse surrounding climate displacees in organisational reports and the relevant academic literature tends to be preoccupied with the threats to security posed by the arrival of climate displacees. The literature very often presents scenarios in which migration will spiral out of control, representing a very real threat. Developed countries in the global North will be directly affected by an 'influx' of climate or indirectly as a result of the consequential changes to regional strategic interests. Such reports are often pre-occupied with the need for a commensurate increase in military expenditure (Elliot 2012: 8–9).

This narrow preoccupation with domestic security concerns, which is often linked to xenophobic and exclusionary impulses (particularly in nations experiencing economic downturn), has the effect of creating what is colloquially referred to as a 'fortress mentality' in which the interests of domestic populations take precedence over those of new arrivals, thereby significantly impacting the human security of climate displacees (Long and Walker 2005: 1). Additionally, in a global geo-political environment where political, religious and other groups suffering persecution are recognised under existing international conventions, the emergence of climate displacees may simply exacerbate intake problems faced by many nations (Fritz 2010: np). The challenges of dealing with displaced populations are immense. Research studies demonstrate that refugee groups face particular problems in terms of high unemployment, housing, education, welfare, language acquisition and dealing with general health and the psychological effects of dislocation (RCA 2011: np). The governmental response to such challenges depends on a variety of factors such as existing entry conditions, adherence to international conventions and whether dislocation occurred through slow or sudden onset climactic factors or whether the movement is intra- or international (Ferris 2011: 7–9).

Emergent realities of population movements

Having identified some of the main challenges facing governments in respect of population movements, it is worth considering specific cases where such developments have already occurred and the issues these raise in terms of governance mechanisms, particularly in respect of interest representation and the context within which deliberation occurs.

Papua New Guinea

An atoll of the Autonomous Region of Bougainville (ARB), part of Papua New Guinea, Carteret Island has become the focus of world attention as a result of the threat posed to bio-diversity and local populations owing to rising sea levels that now inundate its eroding shoreline and plantations. Fifteen hundred people live on this tropical island that has for long provided food and sustenance for its inhabitants who are now at extreme risk because of human-induced climate change. As a result, staple foods such as coconut, banana and seasonal breadfruit have been severely impacted by salination and stunted tree growth occasioned by damaging seawater that has risen more than 10 cm over the past two decades. For the islanders, this has led to tremendous hardship in terms of lack of food and water and severe disruption to agriculture, fisheries and traditional cultural practices. Homes are under constant threat and water, vector and airborne diseases have increased. Additionally, changing climatic conditions have also impacted adversely on sanitation facilities and led to increased malnutrition (PSIDS 2009: 3–9).

The challenges of dealing with population resettlement in an impoverished, conflict-ridden developing country like PNG are formidable. Residents on Carteret Island are familiar with the challenges faced due to food shortages, having experienced such phenomena in the 1990s as a result of overcrowding and competition from international fisheries, resulting in the islanders being designated as economic refugees. Some movement of people occurred to other islands but most of the population remained. However, as noted by the governmental organisation of Pacific Small Island Member States (PSIDS): 'Internal migration both within and between islands places an enormous strain on food, housing, education, health, and water, as recipient communities struggle to accommodate the number of people migrating' (2000: 10–11). The initial planning for the resettlement of the island's inhabitants occurred through the Bougainville Resettlement Initiative (BRI 2008: 2–6). This involved meetings of key stakeholders, including Carteret Islanders, officers from the ABG, Bougainville landowners, AusAid representatives and other non-government organisation representatives. Issues for discussion included a wide range of resettlement issues such as landownership, social and resource planning, identification of resettlement families, preparation for relocation, home construction, planting gardens and culturally appropriate means of exchange (BRI 2008: 2–4).

Carteret Island elders took a lead on the above discussions, which generated considerable tension around planning and resource issues.

Non-government organisations played a key role in assisting the island's inhabitants to engage in negotiations with the AGB and landowners in Bougainville in order to secure ownership of higher land owned by the Catholic Church as well as funds to build 10 new homes. The main problem facing the ABG, and in consequence, the island's inhabitants, is that the former is still grappling with the aftermath of a prolonged civil war in which over 20,000 people were killed. The conflict brought much hardship to Bougainville, the government of which is seeking considerable international aid to address the needs of its 200,000 residents. Frustrated by the slow pace of resettlement, the Carteret Council of Elders in 2006 established a non-profit organisation, Tulele Peisa, to expedite the resettlement process (Komai 2011: np). In 2009, five families were moved to land on Tinputz in Bougainville. The Council has plans to assist the immediate resettlement of a further 80 families with 50 others prepared to go when the opportunity allows. A number of families have already moved of their own volition. Unfortunately, the Council is hampered by lack of funds and has thus far been unable to lay plans for the resettlement of other island communities impacted by climate change. In late 2010, the ABG announced plans to relocate 40 families per year over the next 12 years. The government also announced its intention to upgrade heath, education and other facilities as well as build homes at an estimated cost of over US$7.7 Million (ibid: np). Generally, the process of relocation of islanders has been significantly hampered by both lack of procedural urgency that untimely led to the delay of vital decisions and the lack of funds necessary to pay for relocation infrastructure and allied services. In frustration, elders representing the islanders felt compelled to establish their own collective organisation to pursue their interests through what was a multi-tiered web of organisational arrangements.

Italy

As one of Europe's most southerly counties, Italy has attracted considerable attention in terms of future challenges posed by climate displacees entering the country from North Africa. Such challenges are rendered complex not only by dint of Italy's proximity to the African continent but also by the impact of climate change on the country itself. Faced with significant temperature rises, particularly in the southern provinces, Italy has experienced the re-emergence of many tropical diseases, prolonged and more frequent droughts, with the resulting threat of desertification and accompanying water and food shortages (Kington 2007: np). Competition over increasingly scarce and expensive resources (including and especially water), health and welfare implications, ethnic tensions and increased economic instability are the probable fall-outs of such a scenario (Wolf and Menne 2007: 20–31, Femia and Werrell 2012: np).

The inflow of 25,000 refugees fleeing Tunisia due to political unrest in 2011 – mainly to the island of Lampedusa – offers a glimpse into the governance responses of Italy and other nations when confronted with significant population movements. It has been observed that the response of the Italian government to the inflow of Tunisian refugees was initially to offer them accommodation in (poorly resourced) camps and temporary residence certificates, thereby enabling them to travel to France where many have relatives (Dyer 2011: np).

Faced with mounting public concern, in mid-2011 the Italian government closed the camp on Lampedusa and began forcibly deporting Tunisian refugees back to their country of origin. This was done in an atmosphere of growing public concern over Italy's economic affairs and anxieties about the inflow of people from North Africa. Government policies were tinged with populist rhetoric and although there was consultation with the Tunisian government about the process of return, refugees themselves were subject to top–down governance. This raises important questions over future scenarios in which millions of displaced people from climate-stricken regions seek refuge in southern parts of Europe. As Dyer notes, the recent inflow of Tunisian refugees into Italy is a small fraction of the anticipated numbers predicted to flee North Africa and the Middle East in the wake of climate change. Dyer writes, 'fast-forward 30 years, and assume that the average global temperature is two degrees celsius higher than it was in 1990. That's a reasonable assumption if there is not a drastic cut in global greenhouse gas emissions in the next 10 years' (Dyer 2010 np). Although Italy may be able to feed large refugee populations in the short term, and with the assistance of external organisations, the sheer weight of numbers, not to mention the increased price of imported food stuffs from other countries, will make any long-term intake programme unsustainable. Given the likelihood of millions of people seeking refuge, it is not too difficult to envisage major conflicts over key resources such as food, housing, health services, employment and education (ibid: np).

A key question facing Italy will be the response of other countries in Europe to these massive population inflows. Dyer asserts that most strategists in north European countries are aware of impending population movements, but countries like Germany, France and Poland are unlikely to allow refugees from Mediterranean counties in (ibid: np). These questions and the tensions generated by a comparatively small inflow of 25,000 refugees should be seen as a 'dress-rehearsal' for the future when climate change, rather than political unrest, becomes a major driver of people movements. What became abundantly apparent in this instance is that the interests of refugees became deeply enmeshed in populist politics that led government to treat this group as if they were troublesome cargo needing to be conveniently dispatched in order to placate public opinion and ensure political legitimacy – a response shared to some degree by France as Tunisian refugees flowed into that country (Leduc 2011: np). It is useful at this juncture to consider broadly how the actions of both Italy and PNG in respect to displaced persons compare and contrast with international conventions and processes of good governance. In the wake of the Second World War, the United Nations (UN) on 28 July 1951 evoked the Convention Relating to the Status of Refugees. Under 'General provisions' in Chapter 1, the definition of 'refugee' contained in Article 1 encompassed a range of circumstances in which exiled people in fear of persecution on the grounds of 'race, religion, nationality, membership of a particular social group or political opinion' could seek sanctuary in another country. While not referring or indeed anticipating circumstances resulting in displacement because of climatic conditions, the Convention nonetheless set out various obligations on the part of host nations in terms of the treatment of refugees. Arguably, although climate displacees leave their homelands ostensibly for reasons other than direct persecution it is nonetheless evident, as argued above, that such movements arise in part because of the actions of richer industrialised nations. Thus, while not 'persecuted' as such, displacees remain unable to reside in their countries of residence for reasons not of their making. It is in this sense that the UN Convention has applicability to climate displacees. Articles in the Convention dealing with rights of association (15), access to courts (16) and employment rights (17), housing (21), public education (22), public relief and assistance (23), freedom of movement (26) and transfer of assets (30) indicate provisions that enable refugees to be accorded rights afforded to those displaced from their country of origin (OUNHCHR 1951: np). In the case of Italy, it is apparent that many of the provisions contained in the UN Convention have not been adhered to, for instance in relation to housing and access to public education, as well as many other areas. This was due in part to the politicisation of the presence of large numbers of migrants from North Africa, which encouraged political leaders to expedite the situation by finding ways of moving migrants to other countries. In the case of PNG, although more considered and culturally sensitive in terms of the needs and requirements of displaced persons, the absence of adequate funds to meet the full costs of relocation, as well as the inadequate state of existing infrastructure, meant that significant numbers of people could not be relocated. In terms of the governance mechanisms identified by Cadman – namely interest representation, inclusiveness, equality and resources, accountability and transparency and consultative decision-making and implementation – it is apparent in the case of Italy that the dominant imperative was not to accord with a high level of commitment to such mechanisms but rather to expedite as rapidly as possible the presenting situation in such a way as to ensure political credibility. There is little evidence to suggest that attempts to meet the provisions set out in the UN Convention or the aim of achieving good quality governance in respect of the populations concerned was ever likely to occur so long as the government's actions were being shaped by populist, rather than humanitarian, considerations. In the case of PNG, there is evidence that efforts were made to represent the interests of displaced people through the lead role adopted by elders and the foundation of an organisation to promote the interests of the community as a whole. Certainly, the lack of resources played a significant part in obstructing the implementation of a comprehensive strategy for relocation, and the fact that elders created their own organisation to promote their goals suggests that original decision-making and implementation processes were less than adequate.

Conclusions and recommendations

This chapter has focused on some of the major implications arising from anthropogenic climate change insofar as it impacts all nations, but especially the most economically disadvantaged and poverty stricken. It has been argued that severe and ongoing climatic changes will result in profound alterations to geo-political relations, resulting in mass population movements, and inter- and intra-national conflicts arsing from resource and territorial disputes. The challenges ahead in terms of governance are formidable – as demonstrated in case studies of Italy and Papua New Guinea – not least in seeking to manage resource allocation and avoiding civil strife and violent conflicts.

Effective governance in respect to climate displacees will of course require a differentiated and staggered approach that takes stock of the various circumstances of host nations, the pace of migration and the legal status of refugee relocation. Additionally, the movement of people into other regions or countries brings with it a host of difficulties, not simply to do with the relationship between displacees and host populations, but within these respective populations. Indeed, as an aggregated group embodying differences of class, age, gender, ethnicity, sexual orientation and geographic location, the concept 'population' soon dissolves away from categorical certainty. This point is important when it comes to understanding the immense challenges associated with climate displacees and the likelihood of ensuing conflicts. Above all, it highlights the need to respond to socio-cultural differences and to appreciate their capacity to generate all manners of disruptions between and within nations.

References

Asian Development Bank (ADB). 2012. *Addressing climate change and migration in Asia and the Pacific*. Mandaluyong City, PA: Asian Development Bank.

Bougainville Resettlement Initiative (BRI). 2008. Meeting report, Canberra, December http://displacementsolutions.org/wp-content/uploads/BG.pdf. 11, accessed 15 May 2011.

Cadman, T. 2012. *Evaluating the quality of global governance: a theoretical and analytical perspective*. Earth System Governance working paper, http://www.earthsystemgovernance.org/sites/default/files/publications/files/ESG-WorkingPaper-20_Cadman.pdf, accessed 18 June 2012.

Cadman, T. 2011. *Quality and legitimacy of global governance: case lessons from forestry*. Basingstoke: Palgrave Macmillan.

Commonwealth Scientific and Industrial Research Organization (CSIRO) and Bureau of Metrology (BOM). 2012. *State of the climate 2012*. Canberra: Australian Commonwealth Government.

Development Research Centre on Migration, Globalisation and Poverty, (DRCMGP). 2008. Migration and climate change: How will climate shifts affect migration trends? Briefing No. 12. http://www.migrationdrc.org/publications/briefing_papers/BP12.pdf, accessed 23 May 2012.

Doherty, T. and Clayton, S. 2011. *The psychological impacts of global climate change*, http://climateforce.net/2011/08/07/the-psychological-impacts-of-global-climate-change/, accessed 28 February 2012.

Dunlap, R.E. and McCright, A.M. 2011. Climate change denial: Sources, actors, and strategies. *In:* T. Constance, ed. *Routledge handbook of climate change and society*. London: Taylor & Francis.

Dyer, G. 2011. *Tunisian migration to EU a harbinger of future climate refugees*, http://www.straight.com/article-387847/vancouver/gwynne-dyer-tunisian-migration-eu-harbinger-future-climate-refugees, accessed 24 May 2012.

Elliot, L. 2012. *Climate change and migration in South East Asia: responding to a new human security challenge*. Asian Security Initiative Policy Series Working Paper No. 20.

Elliot, L. 2011. Climate migration: Why it is a human security issue. *Eurasia Review*, March 10, 20, http://www.eurasiareview.com/01032011-climate-migration-why-it-is-a-human-security-issue/, accessed 3 January 2012.

Environmental Justice Foundation. 2008. *No place like home: where next for climate refugees.* London: EJF, http://www.ejfoundation.org/pdf/climate_refugees_final.pdf, accessed 12 January 2012.

Evans, A. 2010. *Resource scarcity, climate change and the risk of violent conflict.* New York: Centre on International Cooperation, http://siteresources.worldbank.org/EXTWDR2011/Resources/6406082-1283882418764/WDR_Background_Paper_Evans.pdf, accessed 15 February 2012.

Femia, F. and Werrell, C. 2012. *Climate change impacts on water and security in southern Europe and neighbouring regions*, http://climateandsecurity.org/2012/03/20/european-commission-climate-change-impacts-on-water-and-security-in-southern-europe-and-neighbouring-regions/, accessed 1 April 2012.

Ferris, E. 2011. Climate change and internal displacement: a contribution to the discussion Prepared for UNHCR Roundtable Bellagio, 22–26 February 2011, http://www.brookings.edu/~/media/Files/rc/papers/2011/0228_cc_displacement_ferris/0228_cc_displacement_ferris.pdf, accessed 24 March 2012.

Flannery, T. 2010. *Here on earth, an argument for hope.* London: Routledge and Kegan Paul.

Friedman, L. 2011. Which nations are most at risk from climate change? *New York Times*, 24 February 2011, http://www.nytimes.com/cwire/2011/02/24/24climatewire-which-nations-are-most-vulnerable-to-climate-95690.html?pagewanted=all, accessed 3 March 2012.

Fritz, C. 2010. *Climate change and migration: sorting through complex issues without the hype*, http://www.migrationinformation.org/feature/display.cfm?ID=773, accessed 18 February 2012.

Global Humanitarian Forum. 2010. *The anatomy of a silent crisis human impact report: climate change.* Geneva: GHF, http://www.bb.undp.org/uploads/file/pdfs/energy_environment/CC%20human%20impact%20report.pdf, accessed 14 January 2012.

Gresser, C. 2011. Climate change: countries plan for unpredictable weather. *Financial Times*, 13 October 2011, http://www.ft.com/cms/s/0/561c5308-ef38-11e0-918b-00144feab49a.html#axzz1yYhDw9mO, accessed 17 January 2012.

Hansen, J. 2010. *Storms of my grandchildren: the truth about the coming climate catastrophe and our last chance to save humanity.* New York: Bloomsbury.

Hamilton, C. 2011. *Requiem for a species: why we resist the truth about climate change.* Sydney: Allen and Unwin.

Intergovernmental Panel on Climate Change (IPCC). 2010. A report of Working Group of the Intergovernmental Panel on Climate Change: Summary for Policy makers, http://www.ipcc.ch/pdf/assessment-report/ar4/wg1/ar4-wg1-spm.pdf, accessed 21 February 2012.

International Organization of Migration (IOM). 2011. *Migration, climate change and environmental degradation: a complex nexus*, http://www.iom.int/jahia/Jahia/complex-nexus, accessed 11 March 2012.

International Organization for Migration (IOM). 2009. *Migration and climate change*, http://www.migrationdrc.org/publications/resource_guides/Migration_and_Climate_Change/MRS-31.pdf, accessed 17 February 2012.

Kington, T. 2007. *Climate change brings malaria back to Italy*, http://www.guardian.co.uk/environment/2007/jan/06/italy.climatechange, accessed 18 February 2012.

Komai, M. 2011. *Sun Come Up, a short film on the world's first climate change refugees receives Oscar nomination,* http://www.pina.com.fj/index.php?p=pacnews&m=read&o=7222338614d66f3a409fb3bc5793f2, accessed 7 January 2013.

Leduc, S. 2011. *Refugees face humanitarian crisis on Paris's doorstep,* http://www.france24.com/en/20110425-humanitarian-crisis-paris%E2%80%99-doorstep, accessed 17 June 2012.

Long, S. and Walker, S. 2005. *Climate refugees: the hidden cost of climate change,* http://www.onlineopinion.com.au/view.asp?article=3569, accessed 3 March 2012.

Lovelock, J. 2009. *The vanishing face of Gaia: a final warning: enjoy it while you can.* London: Allen Lane.

McKibben, B. 2010. *Earth: making a life on a tough new planet.* London: Holt.

Nkrumah, K. 1967. *Neo-colonialism: the last stage of imperialism.* London: Nelson.

Office of the United Nations High Commission for Human Rights. 1951. *Convention relating to the status of refugees,* http://www2.ohchr.org/english/law/refugees.htm, accessed 24 June 2012.

Pacific Small Island Developing States (PSIDS), 2009. *Views on the possible security implications of climate change to be included in the report of the secretary-general to the 64th session of the United Nations General Assembly,* http://www.un.org/esa/dsd/resources/res_pdfs/ga-64/cc-inputs/PSIDS_CCIS.pdf, accessed 12 January 2012.

Park, S. 2011. *Climate change and the risk of statelessness: the situation of low-lying island states,* http://www.unhcr.org/4df9cb0c9.pdf, accessed 12 January 2012.

Powel, S. 2012. *US government study: humans are national security threat to oceans and our planet,* http://www.infowars.com/us-government-study-humans-are-national-security-threat-to-oceans-and-our-planet/, accessed 15 June 2012.

Raleigh, C., Jordan, L. and Salehyan, I. 2009. *Assessing the impact of climate change on migration and conflict,* http://siteresources.worldbank.org/EXTSOCIAL DEVELOPMENT/Resources/SDCCWorkingPaper_MigrationandConflict.pdf, accessed 12 February 2012.

Refugee Council of Australia (RCA). 2011. *Current issues – refugee settlement,* http://www.refugeecouncil.org.au/current/settlement.html., accessed 17 March 2012.

Spratt, D. and Sutton, P. 2008. *Climate code red: the case for emergency action.* Melbourne: Scribe.

Tsuma, W. 2011. *Climate change-conflict nexus: framework for policy oriented action, environment, climate change and conflict. Conflict trends,* 2, http://www.accord.org.za/downloads/ct/ct_2011_2.pdf, accessed 13 June 2012.

UN University 2011. No title, http://www.ehs.unu.edu/index.php/article:130?menu=44, accessed 11 February 2012.

Wolf, T. and Menne, B. 2007. *Environment and health risks from climate change and variability in Italy,* http://www.euro.who.int/__data/assets/pdf_file/0007/95920/E90707.pdf, accessed 11 February 2012.

Wright, H., Kristjanson, P. and Bhatta, G. 2012. *Understanding adaptive capacity: sustainable livelihoods and food security in Coastal Bangladesh.*

13
Migration and Climate Change: Global Governance Regimes and the Incorporation of Climate Change Displacement

Andrea C. Berringer

Introduction

The global governance arrangements for climate change-induced migration are currently at the stage of ad hoc development. Migration due to climate change is still considered a new phenomenon; it has yet to be fully conceptualised let alone integrated into institutional policies. Those affected by climate change, to the extent that migration is an option, would be logically under the current umbrella of environmental migration. However, 'environmental migration' is a loosely defined concept and includes people affected by widely differing circumstances from those affected by typhoons to nuclear accidents, for example. Though this label has been used to describe and effectively assist those affected by sudden impact disasters, those affected by climate change have found little help.

The COP 16 talks in Cancun, Mexico, in December 2010 did provide a small step forward towards governance. The United Nations University was able to negotiate a paragraph into 'The Ad Hoc Working Group on Long-Term Cooperative Action under the Convention' that invites Parties to take specific action nationally to enact: 'Measures to enhance understanding, coordination and cooperation with regard to climate change induced displacement, migration and planned relocation, where appropriate, at national, regional and international levels' (Draft decision -/CP.16). While this subsection, like the agreement, is not legally binding, it does ask individual nations to acknowledge the existence of climate change migration at several levels of governance.

Currently, the governance systems that pertain to different forms of migration only offer specific protections based on international agreements to those who fit into their explicit mandates. Because of this, finding ways to fit climate change-induced migrants into current governance regimes has been a difficult task. Determining which, if any, present institution has the legitimacy to effectively govern this emerging phenomenon is the focus of this chapter. It evaluates three intergovernmental regimes, the United Nations High Commissioner for Refugees (UNHCR), the United Nations Office for the Coordination of Humanitarian Affairs (UNOCHA) and the International Organisation for Migration (IOM). Additionally, it analyses their applicability for governing this aspect of climate change adaptation and provides a number of recommendations to ensure the effective incorporation of climate change-induced migrants into these regimes.

Theorising climate change-induced migration and displacement

Climate policy, in a broad sense, is still concerned with mitigating the effects of climate change and is not yet dealing with the inevitable human effects. There is still much uncertainty as to *how* climate change will play out and because of this, potential policies that can guide human adaptation do not have a permanent 'home'. Several conceptual accounts of climate change are offered here, each of which functions under different governance frameworks.

The first of these is captured under the umbrella of environmental migration. There is general agreement on three causes of environmental migrants: natural disasters and environmental or industrial accidents, planned or unplanned relocation due to development and health-related effects due to inadequate resources to maintain life (Cardy 1994: 3). This definition has been influenced by El-Hinnawi who specified that the first category encompassed temporary displacement because of earthquakes, cyclones or environmental accidents; the second those who are permanently displaced due to anthropogenic changes to a habitat such as a dam or Chernobyl; and the third those who migrate temporarily or permanently because the original habitat can no longer support them (EL-Hinnawi 1985).

Climate change displacement will likely span all categories. As natural disasters such as large and more frequent hurricanes and cyclones occur, additional environmental migrants from the first group are likely. If governments decide to erect improved sea walls or divert water into drought areas, an increase of migrants in the second group is also likely. Finally, if people begin to move due to the inability to sustain their lives and livelihoods, they will fall into the third group. This includes those living on coastlines that are being lost to sea-level rise, or agricultural lands that have been blown away due to desertification. These examples are certainly not exhaustive but offer a glimpse as to the way that climate change can exacerbate known groups of environmental migrants. Dun and Gemenne (2008) argue for a better definition of environmental migration in that it is often difficult to isolate environmental factors from other drivers of migration. Environmental factors are challenging to differentiate from other drivers as they are often underlying and not necessarily seen by those affected by them in the same way in which they are viewed by those studying the event. The way that this ultimately plays out is through environmental damage of human habitats, which initiate a chain of events that affects peoples' lives and livelihoods. The driver is not specifically the environment but its effects. People do not migrate simply because of drought but because they cannot produce food anymore. They do not move because of a cyclone but because the cyclone has eradicated the industry that provided the community jobs.

Climate change can also be conceptualised as a 'disaster'. Disasters also displace human populations and constitute their own phenomena. Disaster-induced displacement includes natural and anthropogenic components but needs to be considered carefully. Not every fire, earthquake, drought, epidemic or industrial accident constitutes a disaster, only those that exceed a society's ability to cope and where external aid is required. Robinson (2003: 9) identifies two types of disasters: natural and anthropogenic and separates them into several subcategories.

In terms of climate change, drivers of migration also overlap with disasters. One could argue that climate change is an anthropogenic industrial disaster and all other damage is subsequent. However, the ensuing catastrophes will be through sudden impacts, slow onset events and possible complex emergencies where epidemic disease may take hold. In short, as a slow onset anthropogenic industrial disaster, it is the most comprehensive one imaginable.

The environmental-anthropogenic disaster is an important intersection. Climate change is often seen as an environmental issue (hence the use of 'environmental migration' to classify the phenomenon at hand), but this distances the changes from their cause, which is anthropogenic industrial pollution. This distinction is important for effective governance; without a proper conception of causation for this type of migration it cannot be adequately governed and may even be placed within a regime where it does not fit. One more important point on the matter is the distinction between migration and displacement, the conception being the difference between voluntary movement and forced migration. As bureaucratic labels, they also affect the quality and legitimacy of governance. The conception being suggested here is that climate change is *forcing* people to move who would not otherwise be doing so; the correlation is 'climate migration' to displacement. However, these terms can easily be used interchangeably. Refugees and victims of natural disasters are usually considered forced migrants because the impetus of their flight was beyond their control, but the slow onset processes of climate change may initiate movement before a typical displacement scenario comes into play. Consequently such distinct categories are blurred. The label chosen to represent this type of migration will inevitably affect where it is to be governed. The following section evaluates three different governance regimes in order to assess which, if any, provides the most legitimate institutional context and quality of governance.

The global regime complex

'Refugee' governance: UNHCR

The loudest arguments for the incorporation of those affected by climate change into existing governance structures have come from those who insist on using the refugee regime; many authors and those affected by climate change use the term 'climate refugee' with the expressed purpose of implicating it. This stems from its international legal mechanism that offers secure and binding protections. The stakeholders pushing this interpretation are activists and those who wish to be able to benefit from international protection.

The modern legal designation of 'refugee' originated in ancient times. A form of hospitality, Mediterranean and Near East civilisations often granted asylum to those fleeing violence. It was an early humanitarian gesture recognising a human desire for liberty and protection (Krenz 1966: 91). This first form of international protection had no special regulations, bureaucratic processes or formalities; it was an informal civil pact of profound importance in that it continues to serve as one of the oldest international norms. Krenz explains this as the 'minimum standards' rule that in short, grants foreigners the same treatment as nationals in cases of conflict (1966: 97).

In the modern era, the 1951 Convention and the 1967 Protocol outlined a specific legal definition that is internationally recognised. It is also a confirmed obligation that guarantees a certain set of rights and privileges. Refugees have the right to seek asylum, cannot be returned to their country of origin and have a protected status unlike economic or environmental migrants (Convention Relating to the Status on Refugees 1951). They also need care and maintenance, reestablishment and legal and political protection (Malin 1947: 448). Refugees, as defined by the UN Convention and Protocol and accepted by the signatories, are a recognised humanitarian commitment. They are internationally justified in their migration and are deemed worthy of assistance.

However, not every desperate situation falls under the protection of the Convention. The Convention's definition of a refugee is structured around the concept of persecution, with the only other clearly identified stipulation being that they must cross an international border.

> [O]wing to well founded fear of persecution for reasons of race, religion, nationality, membership in a particular social group or political opinion is outside the country of his own nationality and is unable or, owing to such fear, is unwilling to avail himself of the protection of that country; or who, not having a nationality and being outside the country of his former habitual residence as a result of such events, is unable to or, owing to such fear, is unwilling to return to it. (Article 1, A. Section 2)

Thus it needs to be established that 'climate refugees' are being persecuted in order to activate this type of migration governance. The general conception of persecution is an individual threat. It can be a threat to a person because of who he/she is; a threat to a person's safety because of an inclusion in a specific group; and in the case of a refugee it is an imminent one. 'Climate refugees' then become the unit of institutional analysis in the case of UNHCR.

Ultimately, climate change and environmental degradation pose problems for the traditional conception of 'refugees'. The earliest migrants will most likely blend in with economic migrants who are already legally and bureaucratically distinguished from refugees; an economic migrant is someone who voluntarily moves rather than is being forced out (Black 2001b). Conventional economic migrants move because of the perceived or real possibility of improving their chances for a better livelihood (Huyck and Bouveir 1983). This group can be difficult to define in that when one is unable to find employment or grow sufficient food, the only option is to migrate – or starve. Not all economic migrants are in a desperate situation, and while many may claim to be 'refugees', few can prove they are victims of persecution. Climate change-induced economic migrants pose a unique challenge to refugee regimes in that the earlier they leave, the less life threatening their situation is. However, their circumstances will only deteriorate further; new economic opportunities will never return. This fact differentiates them from other economic migrants who are, without considerable pressure or distress, starting anew. Consequently, this category of migrant does not appear to offer much comfort to climate change- and environmental degradation-induced migrants and displaced persons.

However, the 1951 Convention and 1967 Protocol should not be dismissed outright, since the associated *Handbook on Procedures and Criteria for Determining Refugee Status* incorporates the concept of refugee on 'cumulative grounds'.

> In addition, an applicant may have been subjected to various measures not in themselves amounting to persecution... in such situations, the various elements involved may, if taken together, produce an effect on the mind of the applicant that can reasonably justify a well-founded fear of persecution on cumulative grounds.... Needless to say, it is not possible to lay down a general rule as to what cumulative reasons can give rise to a valid claim to refugee status. This will necessarily depend on all the circumstances, including geographical, historical, and ethnological context. (Interpretation of terms, Section b, Number 53)

Cumulative grounds can include the ways in which climate change will affect the lives of many: its complications, chain reactions and refugee-causing catastrophes. Global climate change will impact a region's ability to produce agricultural goods, will expose more people to floods and drought and threaten the integrity of certain island chains. Arable land and current sources of drinking water may have new geo-political owners. This can easily create struggles for power and incite violence. Martin argues there is a growing concern that scarcity-induced insecurities can contribute to the amplification of the perceived significance of ethnic differences (2005: 330). This may be linked to the progression of events by which people affected by climate change become legal refugees. Equally, Black explains Thomas Homer-Dixon's connections between environmental scarcity and conflict (Black 2001: 2–3). While, on the one hand, Black rejects that environmental scarcity directly leads to conflicts between states, this is a complicated process that can disrupt the public order in several ways (ibid: 4); on the other hand, Homer-Dixon advocates that scarcity initiates large population movements that in turn causes identity conflict and disrupts social institutions leading to wars based on depravation (ibid: 13). Most recently, a United Nations University Institute for Environment and Human Security report based on case studies across refugee camps in Africa revealed the opposite – that pre-existing conflict exacerbated the difficulties from climate variability (Afifi et al. 2012: 46).

Whether 'climate refugees' benefit from the protections of the Convention and its Protocol remains to be seen. There is no clear way that the term 'persecution' can accurately describe what is happening to them, unless a causal link between climate change and human agency can be established in international law. There may be a case to be made if cumulative grounds are considered, but this is also very difficult as climate change is generally slow to take effect. Individuals would have to cite many years' worth of evidence that led to the current need to flee. This would be difficult to prove and also relies on the fact that affected individuals understand that their current misfortunes are causally linked to climate change.

As previously mentioned, what little discussion that has taken place on this issue has been around extending the refugee regime. This regime has already been stressed by pressure to extend its mandate for other reasons. The main source of pressure has been from advocates for Internally Displaced Persons (IDPs). Many conflicts internally displace people who, if they had crossed an international border, would receive assistance as Convention refugees. Expanding its existing mandate could threaten the protections that they provide to refugees in dire need of assistance. UNHCR's mandate is often seen as not flexible enough to deal with more varied types of displacement. As a unit, it has been reluctant to engage in the debate for including environmental migration as a part of its original mandate because of the view that it would dilute current protections, pose a greater burden on receiving states who may be reluctant to accept them as refugees and that environmental refugees currently do not fit the refugee definition under its mandate (Urosevic 2009). However, after the Indian Ocean tsunami of 2005, it became interested in expanding its mandate to cover natural disasters (Long 2010: 12). Former President Nasheed of the Maldives and his UN delegation have already suggested that the United Nations take a vote on expanding the UNHCR mandate to include those affected by climate change (Biermann and Boas 2008: 11). It has not garnered the support necessary to open discussions. A vote to extend the mandate could also be very difficult to attain since states that are reluctant to take in more refugees may vote against it. Considering the global economic situation

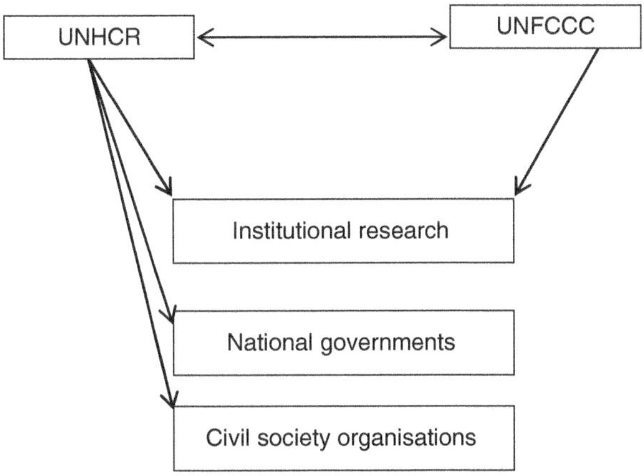

Figure 13.1 UNHCR organisational relationships

as well as issues of terrorism, there is a chance that a vote would be highly unpopular and counterproductive, making it more difficult to take up this issue a second time.

Cadman's conceptual model of contemporary governance points out that collaborative structures and processes contribute considerably to substantive outcomes and ultimately legitimacy (2011: 5). Since the interests of those affected by climate change are not currently included in the UNHCR mandate, can the 'refugee' regime be seen as the appropriate institutional venue to govern such a phenomenon? It is necessary to further evaluate other regimes that deal with migration. By exploring more policy instruments a clearer picture emerges of all of the available possibilities for including those displaced by climate change. This is where IOM and OCHA become important. The subsequent discussions indicate that these are more inherently pre-disposed to combatting the emerging problem of climate change-related migrants and displaced persons than UNHCR (Figure 13.1).

'Migration' governance: IOM

Migration is the story of humanity. From the earliest hominids to modern peoples, history can be understood through human migration; it is almost genetic. Some of the earliest stories of origin come from the Rift Valley in Western Africa and demonstrate humanity's amazing capacity for migration, travelling to every corner of the globe to inhabit desert, ice and small islands in the sea. Humans have never remained static – which is the point often lost in many modern accounts of migration policy. Borders are a new development, and even though globalisation has implied a free flow of information, goods and capital, it has restricted the natural flow of people.

Unlike the refugee regime, the migration regime is not supported by a legally binding treaty. Governance in this area is based on international agreements by its primary stakeholders, which are individual states facilitated through the IOM. With its 146 Member States, it is the intergovernmental organisation that promotes humane and orderly migration for all. Under this mandate, the addition of climate migration is an easy fit. The organisation facilitates migration through providing advice and services to both governments and migrants. It originated to facilitate the physical movement of the refugees that UNHCR would legally protect. This includes transportation, provisions and logistics. IOM orchestrates many projects that are proposed by its Member States. Its mandate is broad and has continued growing as a 'soft' governance structure, that is, one that is voluntary and not bound by intergovernmental conventions or protocols.

IOM acknowledges that climate change will trigger additional movements of people within and between borders and there are difficulties with disentangling the environment from other drivers of migration and displacement (IOM 2011: 2). Its usual activities do not seek to contain peoples but to assist in providing not only short- but also long-term durable solutions. This points favourably to its legitimacy at least with regard to its implementation capacity (Cadman 2011: 15). In terms of climate change, IOM's current activities include disaster risk reduction and capacity-building projects (IOM 2011: 3). However, while these are important components for future adaptation, there is at present nothing substantive concerning the imminent requirement for relocation projects for those people whose homes will eventually become uninhabitable.

IOM would have much less of an issue with being seen as generally legitimate; its core function is to facilitate the movement of people between and across borders and is not hindered by the same legal definitions of UNHCR – requiring persecution and the crossing of an international border. However, it must overcome several serious weaknesses in order to legitimately govern climate-induced migration. As an organisation, IOM is state-centric in that its functions are funded and decided on by its members. To assist those being displaced by climate change would require substantial resources and, in turn, this means having stakeholders included, who represent this interest, are willing to act

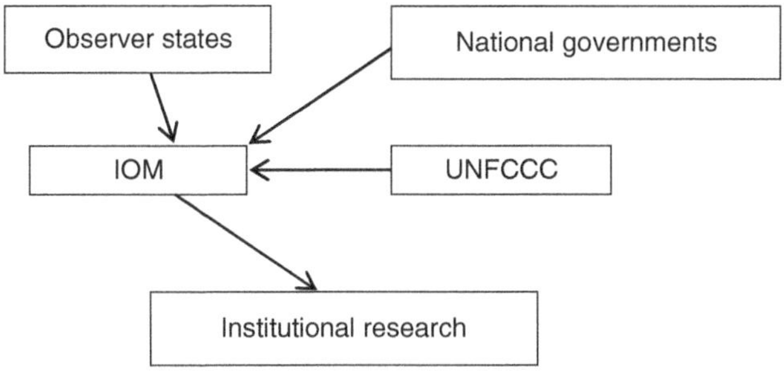

Figure 13.2 IOM organisational relationships

and have the access and weight to do so (Cadman 2011: 12–13). In terms of access, IOM has less Member States than the UN-based governance bodies, and none of those in the Alliance of Small Islands States (AOSIS) who are most vulnerable to climate change – and who are already feeling its effects – are not members. Even if these states were members, they are small, with limited economic resources; weight is not in their favour. This alone does not necessarily keep IOM from being legitimate, but this fact can significantly limit capacity towards this issue. Finally, IOM only facilitates *physical* movement, as per its constitution. This means that it has no obligation to be otherwise responsible for what happens to these migrants when they reach their destination or in helping them find one. Its deliberative processes do not account for the full range of needs that those becoming displaced by climate change will need; thus while the legitimacy of IOM appears amenable at first glance, its processes and structure may not necessarily lend it to be so (Figure 13.2).

'Humanitarian' governance: OCHA

Unlike the historical refugee tradition, the humanitarian regime, with its organised set of institutional arrangements, is a modern phenomenon. The idea of doing something altruistic or philanthropic is not new, but the organisation of collective action to alleviate suffering is. Its origins can be traced to the creation of International Committee for the Red Cross and the American Red Cross in the mid-nineteenth centuries, but it is important to note that aid of this sort in the beginning was directed solely at medical need. Even into the mid-twentieth century, it was still the belief and practice that the responsibility, interest and capacity to assist individuals in a disaster situation lay with the national government of the affected area (Leebaw 2007: 224).

In the post-Cold War era, the humanitarian regime grew exponentially due to several specific environmental factors. First, geo-political shifts at the end of the war increased demand for humanitarian action; without Soviet- or US-sponsored aid, unstable domestic situations threatened to become large emergencies (Barnett 2005: 726). Additionally, state spending on humanitarian aid increased dramatically as nations began to show an interest in utilising such aid in connection with political goals; it was also seen as a rationale for regime change (Leebaw 2007: 226). Private contributions also increased, but not nearly as fast as official assistance with the United States as the lead donor. Political motives fuelled this increase in giving and conditions were often placed upon such aid. Finally, there was also a change in the legal environment; the concept of state sovereignty was becoming conditional based on accepted behaviour to ones' own people (Barnett 2005: 725).

The UNOCHA was established in 1991 to fill in the gaps of international humanitarian assistance. OCHA's mission is to: coordinate and mobilise humanitarian action in an effective and principled manner in partnership with national and international actors in order to alleviate human suffering in disasters and emergencies; advocate for the rights of people in need; promote preparedness and prevention; and facilitate sustainable solutions (UNOCHA n.d.). The stakeholders involved in its development were not only individual states but also UNHCR, which was under pressure to handle these types of situations. OCHA represents a humanitarian regime driven by human sympathy and the obligation to better the human condition. OCHA also assists in circumstances where there is internal displacement due to disaster. Its mandate is different from UNCHR and is not restricted by the international movement of disaster victims or the label of 'refugee'. OCHA's main goal is to make emergency response more effective. However, its mandate does rely on the effected nations' responsibility to its own people.

> Responsibility for people affected by emergency lies – first and foremost – with their respective states. States in need are expected to facilitate the work of responding organizations. Humanitarian assistance must be linked to the humanitarian principles of humanity, neutrality and impartiality.
>
> (UN General Assembly resolution 42/182)

Although OCHA organises many responses, it does not take direct responsibility for those uprooted by disasters. To date, OCHA has

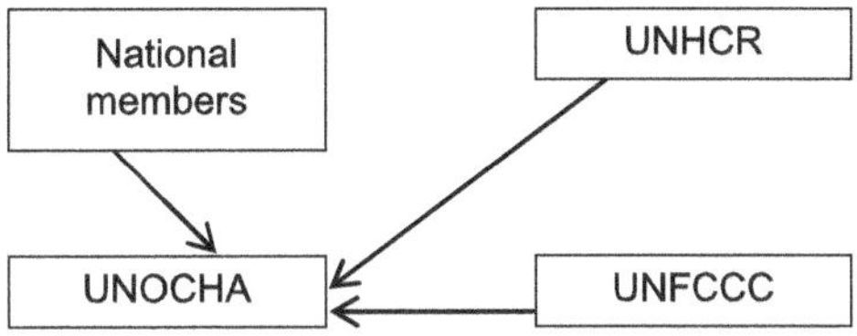

Figure 13.3 UNOCHA organisational relationships

not yet provided a framework to specifically assist those who will potentially be uprooted. It does offer a section on its website about climate change and coping with the humanitarian response. It explains the threat of climate change, who is most vulnerable to it and provides some informational statistics on the matter, for example that 70 per cent of all disasters today are climate related, and between 1998 and 2007, 2.2 billion were affected by climate disasters (UNOCHA n.d.). Even though OCHA identifies climate change as the underlying driver of today's disasters, its policy suggestions are geared towards adaptation and mitigation – not yet tackling the emerging problem of climate-related displacement itself (Figure 13.3).

Which regime most fits climate-induced migration?

It is important to pause and connect back to the theme of this volume. The predominant international and intergovernmental discourse around climate change-induced migration and displacement has been couched within the international refugee regime. This is an important and influential regime, but forcing these emerging issues into this regime will not necessarily result in effective governance. Dover and Hezri (2010) explain this paradox of institutional fit. Long-lived institutions include a balance of predictability and adaptability, but are prisoners of history as they typically embody past understandings, not present or future ones. UNHCR was initiated for a very specific purpose in a very unique moment in time. Considering this fact, it has still been an amazingly effective regime; this is why the term 'refugee' continues to be used to describe desperate situations. Beyond its descriptive appeal, it implies an international mechanism to address the issue.

IOM will need the money, support and political will of its large industrial members to be legitimate. This is not impossible but is still a long shot in this current political climate. Most importantly, using the migration regime can be a double-edged sword. While there is the need to focus this issue on human movement itself, it can happen at the expense of its cause. Using 'migration' may not lend itself to policy solutions that consider that what is happening is 'displacement'.

The only international governance system that has a chance to be seen as legitimate is OCHA. It currently assists those that are at least temporarily living outside of their habitual residence due to generalised violence or the effects of a natural disaster; thus, OCHA is no stranger to dealing with many forms of migration and displacement. If the adverse effects of climate change on human populations can be considered a humanitarian issue, this opens the door for legitimacy through this institution. The effects of climate change are complicated in that they exacerbate many other societal issues; OCHA works at the intersection of these issues when people are at their most vulnerable. Its mandate is broad enough to accommodate many conceptions of this phenomenon (environmental migrants and/or displacees and disaster victims of all types), and because it innately assists those in many vulnerable positions for extended amounts of time its legitimacy for transport and transition is assured.

Conclusions and recommendations

Despite the arguments that advance above, for many, the way forward still lies with UNHCR. No matter its reluctance to take on this new migration flow of displacees or the lack of legal recognition for the 'climate refugee', it has a unique character and has been known to attend to IDP's when it has the ability. Whether it takes the lead will depend on two conditions: the ability of UNHCR to fully protect Convention refugees; and to adequately fund and staff any new expansion. UNHCR needs to be able to attend to its original mandate *while* expanding it. However, because this will most likely involve a new treaty in order be fully funded. Simply asking current signatories for more money to expand its mandate could come up against political pressure from industrialised nations. These nations already contribute to UNHCR and many also contribute to adaptation funds. Apart from the UN, they are also the larger contributors to disaster relief and development projects, which means they are likely to cry giving fatigue arguing funding climate migration is a charitable donation instead of a continued responsibility.

A new international treaty that recognises migration as an appropriate adaptation measure and that outlines and facilitates this movement is necessary. This should be either through OCHA or IOM. The purpose of this treaty would be to assist migration and resettlement, not provide indefinite protections from persecution. There should be a variety of migration statuses that allow those from the most vulnerable hot spots to 'jump' to the front of the line to immigrate into other signatory nations before those desiring to move under less urgent circumstances. This will serve to reduce the confusion with economic migrants or Convention refugees. A treaty of this sort would recognise the specific interests of potential migrants and could be integrated into existing national immigration policies that would strengthen its legitimacy.

A treaty through IOM associates this movement of people with other reasons for orderly relocation. IOM's mandate fits the circumstances of those affected by climate change more clearly than OCHA and disassociates the frame from aid/charity, which may help to combat donor fatigue. However, IOM will need to be significantly expanded and allocated much more resources in order to adequately address this situation legitimately.

While the preceding options stem from two different governance regimes (humanitarian and migration), each offers a better 'fit' than that of UNHCR (the refugee regime). But neither option will materialise in an easy manner. There is a major power differential between the stakeholders in any agreement of this type: the global North versus the global South and the rich versus the poor. There is still little incentive for those most responsible for this growing phenomenon to prevent or facilitate it in an orderly fashion. Most movement is occurring in poorer areas where individuals' livelihoods are more closely tied to nature. For many nations, this issue is still out of sight and out of mind. That said there is still the potential for 'good' governance to emerge. The acknowledgement of the phenomenon of migration due to climate change as acknowledged in the Cancun agreement was a big step forward, but nothing more materialised from the Durban talks. For those currently affected, the COP negotiations have provided a platform to be heard and acknowledged by those who are presently removed from the situation. It is the start of a deliberative process from a participatory structure that, if it can produce substantive results, will not only be far more legitimate but will also deliver quality of governance.

References

Afifi, T., Govil, R., Sakdapolrak, P. and Warner, K. 2012. *Climate change, vulnerability and human mobility: perspectives of refugees from the east and horn of Africa.* No location: UNU Institute for Environmental and Human Security.

Barnett, M. 2005. Humanitarianism transformed. *Perspectives on politics*, 3, 723–740.

Biermann, F. and Boas, I. 2008. Protecting climate refugees: the case for a global protocol. *Environment: science and policy for sustainable development*, 6, 8–17.

Black, R. 2001a. Environmental refugees: myth or reality? *New issues in refugee research*, working paper 34.

———. 2001b. Fifty years of refugee studies: from theory to policy. *International migration review*, 35 (1), 57–78.

Cadman, T. 2011. *Quality and legitimacy of global governance: case lessons from forestry*. Basingstoke: Palgrave Macmillan.

Cardy, W.F. 1994. Environment and forced migration, a review. Sommerville College, University of Oxford, United Kingdom: Fourth International Research and Advisory Panel Conference, January 1994.

Dovers, S.R. and Hezri, A.A. 2010. Institutions and policy processes: the means to the ends of adaptation. *Wiley interdisciplinary reviews: climate change*, 1 (2), 212–231.

Dun, O. and Gemenne, F. 2008. Defining 'environmental migration'. *Forced migration review*, 31, 10–11, http://www.fmreview.org/sites/fmr/files/FMRdownloads/en/FMRpdfs/FMR31/FMR31.pdf, accessed 15 July 2012.

El-Hinnawi, E. 1985. *Environmental refugees*. Nairobi: UNEP.

Huyck, E.E. and Bouvier, L.F. 1983. The demography of refugees. *The annals of the American academy of political and social science*, 467, 39–61.

IOM. 2011. *World migration report 2011: communicating effectively about migration*. Geneva: International Organization for Migration.

Krenz, F.E. 1966. The refugee as a subject of international law. *ICLQ*, 15 (1), 90–116.

Leebaw, B. 2007. The politics of impartial activism: humanitarianism and human rights. *Perspectives on politics*, 5, 223–239.

Long, K. 2010. *Forced migration research and policy: overview of current trends and future directions*. Oxford: Refugee Studies Centre.

Malin, P.M. 1947. The refugee: a problem for international organization. *International organization*, 1 (3), 443–459.

Martin, A. 2005. Environmental conflict between refugee and host communities. *Journal of peace research*, 42 (3), 329–346.

Robinson, W. 2003. *Risks and rights: the causes, consequences, and challenges of development-induced displacement*. Washington DC: The Brookings Institute-SIAS Project on Internal Displacement.

UNHCR. 1951. *The convention relating to the status of refugees*. No location: UNHCR.

UNOCHA, n.d. *Climate change – humanitarian impact*, http://www.unocha.org/what-we-do/advocacy/thematiccampaigns/climate-change/humanitarian-impact, accessed 7 January 2013

Urosevic, N. 2009. Environmental 'refugees': should the UNHCR enlarge its mandate to include environmental migrants? *Undercurrent*, 6 (3), 27–34.

14
The ICLEI Cities for Climate Protection Programme: Local Government Networks in Urban Climate Governance

Heather Zeppel

Introduction

This chapter looks at the role of transnational local government (LG) networks in urban climate governance, where the term urban includes cities, towns and other metropolitan municipalities. Local authorities are recognised as one of the nine major groups in Agenda 21 and included in the UN Commission of Sustainable Development (UN 2009). LG was officially recognised as a governmental stakeholder at the 2010 COP 16 climate summit and United Nations (UN) meetings now include LG interests in global climate agreements. LGs have been actively engaged in climate change policies since the early 1990s. Cities and other municipal authorities have prepared climate change action plans and strategies addressing local mitigation and adaptation measures. Many LG authorities, particularly cities, are members of national and transnational municipal networks addressing climate change such as International Council for Local Environmental Initiatives (ICLEI), The Covenant of Mayors (Europe), US Mayors Climate Protection Agreement (USA), and C40 Cities (global megacities). These LG climate networks focus on informing members about best practice methods, learning about climate change issues and representing municipalities at national, continental and international levels and at climate forums. Hence, 'networking within and across municipal [and national] borders has become commonplace in local and regional climate governance' (Granberg and Elander 2007: 545). LG authorities are key stakeholders in responding to climate change as they control vital areas and assets that affect greenhouse gas (GHG) emissions such as land-use planning, building codes and standards, transportation, energy infrastructure, waste services and water or wastewater utilities. Climate change policy though is still mainly a voluntary task for local authorities, with climate governance focused on citizen education programmes and corporate mitigation of emissions rather than regulations (Gore 2010: 27–46). Municipal climate action thus represents 'a reconfiguration and rescaling of climate governance' at the local level, mainly driven by civic society initiatives through urban climate networks (ibid: 42). Following Cadman (2011), this chapter assesses the governance of ICLEI's Cities for Climate Protection (CCP) Programme, evaluating the structures and processes associated with interest representation, organisational responsibility, decision-making and implementation of climate policy by LG.

Cities and climate change

Cities and larger municipalities are key stakeholders in climate change policies and actions, due to their size, role and functions (Betsill and Bulkeley 2007: 447–456; Bulkeley and Betsill 2003). Cities are high consumers of energy and producers of waste and GHG emissions. An estimated 30–40 per cent of human-caused GHG emissions emanate from within cities (Satterthwaite 2008: 539–549). Over 50 per cent of global populations now live in cities, growing to 60 per cent by 2030. Cities consume more than two-thirds of the world's energy and account for 75 per cent of global GHG emissions (C40 Cities 2011). The citywide GHG emissions for the largest 40 cities ('C40') collectively exceed the individual annual emissions of 167 United Nations Framework Convention on Climate Change (UNFCCC) countries (Arikan 2011). Local municipalities thus have a significant role to play in reducing emissions, due to their control of energy, transport, land use planning, building codes, waste management and community education. LG involvement in sustainable development through Local Agenda 21 since the 1992 Rio Earth Summit has facilitated climate mitigation actions. LG can also support action on climate change by implementing local demonstration projects and lobbying state or national governments on climate policy. Some 19 global cities are members of UNEP's Climate Neutral Network to reduce urban GHG emissions (UNEP 2012). LG experience and expertise in managing environmental impacts also assists with local measures to reduce emissions. Municipal climate actions are thus an essential part of moving towards a lower carbon society and reducing GHG emissions (Chavez and Ramaswami 2011: 471–482; Coenen and Menkveld 2002: 107–125; Storey et al. 2012). The key role of cities in reducing urban GHG emissions is now recognised in state and national climate strategies and reduction targets. Cities are also taking independent action on sustainability and climate change in response to global efforts and local community concerns (Anguelovski and Carmin 2011: 169–175). These urban climate change adaptation and mitigation strategies need to be complementary and avoid potential conflicts between energy use and emissions reduction (Hamin and Gurran 2009: 238–245). Transnational LG climate networks such as the CCP Programme managed by ICLEI play a key role in disseminating climate policy, best practice and advocacy for urban climate action.

Recent OECD reports address cities, climate change and multi-level governance, and competitive cities and climate change (Corfee-Morlot et al. 2009; Kamal-Chaoui and Robert 2009). A World Bank report, *Cities and climate change: an urgent agenda*, reviewed climate impacts on cities, urban GHG emissions, the benefits of urban climate action and financial support or climate partnerships for cities (World Bank 2010). Another report addresses climate action in megacities, and other reports focus on disclosing urban GHG emissions of large cities (C40 Cities 2011; Carbon Disclosure Project 2010; 2011). In North America, cities are leaders in climate change actions despite national climate policies in the United States and Canada not supporting Kyoto Protocol (KP) targets (Gore and Robinson 2009: 137–158). The content and effectiveness of climate action plans have been reviewed for US cities (Bassett and Shandas 2010: 463–475; Boswell et al. 2010: 451–461; Wheeler 2008: 481–491). Wheeler found these climate change plans for US cities mainly focused on actions with energy and cost savings while one-third did not quantify GHG reduction goals. The CCP programme requires cities to prepare a climate change action plan, but emissions goals vary, while many climate actions are voluntary, not implemented or lack resources (Wheeler 2008: 481–491).

Local government and climate change: institutional analysis of ICLEI and CCP

ICLEI was founded in 1990 as a global network for LG and cities achieving environmental sustainability through local actions. Since 2003, it has been known as ICLEI: Local Governments for Sustainability. Based in Toronto (Canada) with a world secretariat in Bonn (Germany),

ICLEI has regional divisions located in Africa, Canada, Europe, Japan, Korea, Latin America and Caribbean, Mexico, Oceania (Australia/New Zealand), South East Asia, South Asia and the United States. ICLEI has over 1,220 LG members and associations in 80 countries representing half a billion people. The LG members of ICLEI include 12 mega-cities, 100 super-cities, 450 large cities/urban regions and some 650 small- and medium-sized cities and towns. ICLEI is governed by a global council with a minimum of 27 and a maximum of 45 seats and comprises members of the nine regional executive committees. ICLEI also manages the World Mayors Council on Climate Change and the Bonn Centre for Local Climate Action and Reporting. Other ICLEI programmes or 'flagship brands' include the following: Future City Leaders, Cities Climate Registry, Green Climate Cities, EcoBudget, Local Action for Biodiversity, Procura Plus, Local Renewables, EcoMobility and Resilient Cities (ICLEI 2012a). ICLEI USA has over 600 LG members covering 25 per cent of the US population. ICLEI USA also released a Local Government Operations Protocol to measure and report municipal GHG emissions in 2008 and a community GHG emissions protocol in 2011 (Ewing-Thiel and Manarolla 2011: 371–375). In 2012, ICLEI launched a Global Protocol for Community (GPC)-Scale GHG Emissions at the UNFCCC Bonn Climate Conference.

The CCP campaign, established by ICLEI in 1993, has focused on lowering GHG emissions at the LG level (ICLEI 1997; 2006; 2010a). ICLEI also partners with other global networks for climate action such as the UNEP Climate Neutral Network, World Mayors Council on Climate Change (WMCCC), Local Government Climate Roadmap, C40 Cities and Bonn Center for Local Climate Action and Reporting with its Cities Climate Registry. ICLEI has represented LG at the UNFCCC Conference of Parties (COP) global climate summits since 1995, and in 2009 became the first LG observer organisation at the IPCC. ICLEI has led the Local Government Climate Roadmap process at UNFCCC meetings since 2007. A World Mayors and Local Governments Climate Protection Agreement was launched at COP 13 in Bali in 2007, advocating for a 60 per cent global reduction of GHG emissions by 2050. ICLEI is part of a coalition of LG global networks (that is, WMCCC, United Cities and Local Government, and C40 Cities Climate Leadership Group) advocating for greater recognition of LG and setting GHG targets after the KP ends in late 2012. A Cities Climate Registry was launched at the World Mayors Summit on Climate in 2010, with public reporting of climate commitments, actions and GHG emissions by cities (Arikan 2011). As on May 2012, the Cities Climate Registry listed climate and energy commitments and emissions information of 164 cities from 21 countries (ICLEI 2012b). After 15 years of lobbying by ICLEI to include cities in global climate negotiations, LG was officially recognised as a governmental stakeholder at the 2010 COP 16 climate summit. COP meetings now formally include LG interests in global climate agreements, through advocacy by a local government and municipal authorities (LGMA) constituency to the UNFCCC (Okereke et al. 2009: 58–78; Rabe 2007: 423–444). Local authorities are recognised as one of the nine major groups in Agenda 21 and included in the UN Commission of Sustainable Development (UN 2009). The involvement of ICLEI in global climate negotiations illustrates 'A key concern [of governance] is processes of networking and partnerships' (Stoker 2000: 93) (Figure 14.1).

Other international LG networks have implemented climate change initiatives. In 2011, the United Cities and LG group presented submissions at COP 17 in Durban, while Metropolis has included a 'Cities for Climate Change' agenda as part of their Commission 1: Ecological Region since 2005 (Metropolis 2009; 2011; United Cities and Local Government 2011). The US Conference of Mayors launched a US Mayors' Climate Protection Agreement in 2005 and a Climate Protection Centre in 2007. A similar pledge for mitigation actions was endorsed by The Covenant of Mayors for Europe in 2008, where 4,050 signatories aim to exceed the European Union's objective of a 20 per cent reduction in emissions by 2020 by implementing Sustainable Energy Action Plans

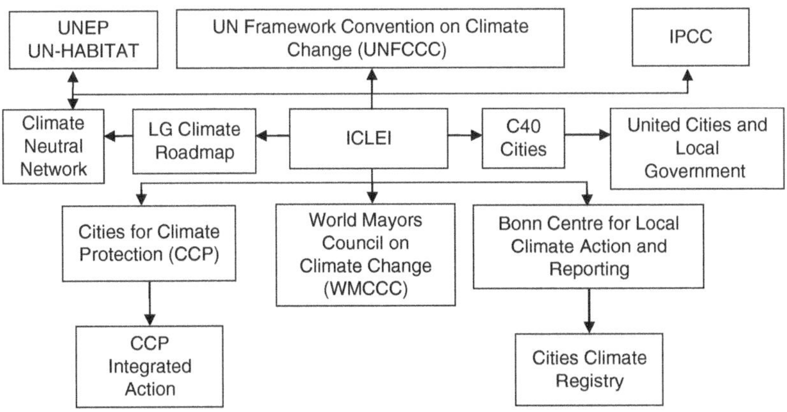

Figure 14.1 Institutional relations between LG climate initiatives and the global climate regime

in their area (EU Mayors 2012). The European Commission also ran a specific LG Action programme on climate change (2009–2011), while the United Kingdom implemented a Low Carbon Cities programme in 2008. In the United Kingdom, 90 per cent of local areas had signed up to climate change National Indicators, abolished by the British government in October 2010 (Ross 2011). Climate change goals are now negotiated in Local Area Agreements as part of carbon reduction commitments in the United Kingdom (Pearce and Cooper 2011: 199–217). The C40 Cities Climate Leadership Group, representing global megacities, has partnered with the Clinton Climate Initiative since 2006 to implement mitigation programmes, the World Bank Institute on carbon finance for C40 cities and the Carbon Disclosure Project (CDP) to report on the GHG emissions of C40 cities (C40 Cities 2011; Carbon Disclosure Project 2010; 2011). The cities and climate change agenda of the World Bank includes the Eco2 Cities Programme, GHG standard, carbon finance for cities, Energy Efficient Cities Initiative and Global Cities Indicators Programme (World Bank 2010). LG declarations on climate change from 1993 to 2011 have endorsed and supported the role of municipalities in taking direct action on climate change and reducing GHG emissions. ICLEI has led LG involvement in these global climate negotiations and declarations, in partnership with other LG climate networks and through CCP.

Cities for Climate Protection

The CCP campaign was established by ICLEI in 1993. Globally, over 800 municipalities have participated in CCP's climate network. There is an allied ICLEI Partners for Climate Protection programme in Canada and ICLEI Communities for Climate Protection programme in New Zealand. The CCP programme focused on lowering GHG emissions at the LG/city scale by municipal authorities and their communities. ICLEI's CCP campaign was an outcome of the first Municipal Leaders Summit on Climate Change in New York in 1993, following the adoption of the UNFCCC at the Rio Earth Summit in 1992. ICLEI had previously implemented an Urban CO_2 Reduction campaign in 1991 involving local climate action in 14 cities across Europe, the United States and Canada. The CCP programme was adopted by cities across Europe, North America and Australia in the early to mid-1990s, followed by cities in Latin America, South East Asia, India and South Africa from 1999 to 2002 (ICLEI 1997; 2005; 2006; 2010a; 2010b; 2011). Australia participated in the CCP programme from 1997 to 2009, and New Zealand from 2004 to 2009, with a total of 272 CCP members in these countries (Hoff 2010). In Australia, federal government funding for CCP ended in 2009, and CCP Partners now includes just 60 LGs in Australia and New Zealand (ICLEI Oceania 2011).

The CPP programme provided a tool for municipal authorities to initiate climate mitigation and adaptation actions at the local level. Municipalities joining CCP pledge to reduce GHG emissions arising from (1) the operations of LG and (2) from their local communities. They commit to an emissions reduction target (from a base year to a target year), plan their carbon mitigation actions (that is, energy efficiency, renewable energy and waste management), prepare and implement a local climate action plan and monitor reductions in GHG emissions due to mitigation actions. The CCP methodology and tools assist municipalities to calculate their GHG emissions, set emissions targets, reduce GHG emissions and then monitor, measure and report their performance on mitigation goals. CCP is based on five milestones with LG selecting their desired milestone. The benefits of the CCP programme to LG were GHG reductions, financial savings, greater awareness of climate change in councils and communities, local leadership and integrated programmes on climate change with key stakeholders and a network of CCP councils.

The CCP programme is administered by a national ICLEI office in each country. These provide technical advice, training workshops and support to participating municipalities. Betsill and Bulkeley (2004) reviewed CCP as a transnational network involved in policy learning about climate change mitigation at the LG level. They found municipalities engaged with the CCP programme mainly due to the financial and political resources it offered and the legitimacy it provided for local climate protection actions. Different LG actors involved in CCP also had varied interpretations of and meanings about local climate change policies (Betsill and Bulkeley 2004: 471–493; Betsill and Bulkeley 2006: 141–159). Bulkeley and Betsill also evaluated the impact of the CCP programme on six cities, in the United Kingdom (3), the United States (2) and Australia (1). They found the CCP programme had varied impact on the implementation of local climate protection policies by these cities due to: (1) a committed individual with institutional support, (2) availability of climate funding, (3) level of local power over city functions, (4) framing of climate change (for example economic objectives) and (5) political will (Bulkeley and Betsill 2003).

Governance analysis of Cities for Climate Protection

This analysis of CCP as a transnational climate network focuses on CCP's structures and processes following the approach to assessing governance quality laid out in the Introduction to this volume. Key criteria for assessing CCP include interest representation, organisational responsibility, decision-making and implementation (See Table 14.1). Information has been sources from academic literature, ICLEI and CCP websites, and discussions with previous CCP programme members.

Interest representation

ICLEI manages CCP as a global climate network for LG climate efforts, administered through regional offices and partnerships (for example Partners for Climate Protection, Canada; Communities for Climate Protection, New Zealand). CCP represents LG interests in climate change mitigation, such as benefits from reduced GHG emissions, cost savings and city investments in clean technology. CCP participants include cities and towns in both developed and developing countries, with CCP adopted by some 25 municipalities in South East Asia and India (Yienger et al. 2002). In terms of inclusiveness, ICLEI and the CCP programme is driven by the climate agenda of developed countries and CCP members are mainly larger cities rather than smaller municipalities (Krause 2011: 45–60). The global North dominates other urban climate networks' governance arrangements (such as Metropolis and C40 Cities) (Bouteligier 2011). In terms of equality, local communities also have no input into CCP and are regarded as passive recipients of LG climate change measures (Slocum 2004: 763–782). While community emissions

Table 14.1 Criterion-level framework for the assessment of governance quality applied to CCP, following Cadman (2011)

Criterion	Indicator	CCP programme
Interest representation	Inclusiveness	CCP inclusive of LG interests
	Equality	CCP treats all LG interests equally
	Resources	Level of CCP resources for LG
Organisational responsibility	Accountability	CCP accountable in dealings with LG
	Transparency	CCP transparent in dealings with LG
Decision-making	Democracy	CCP processes are democratic
	Agreement	Making of CCP agreements effective
	Dispute settlement	Effective settling of disputes in CCP
Implementation	Behavioural change	CCP changing LG behaviour
	Problem solving	CCP helping to address climate change
	Durability	Durability of CCP programme

are included in CCP, these are subsidiary to municipal abatement efforts, mainly due to challenges with community-wide social change programmes and actions. CCP mainly focuses on LG initiatives in energy, waste and transport and the business case for reducing emissions, rather than equitable community benefits. Smaller LG authorities are less likely to participate in CCP due to cost and resourcing issues.

Municipalities engaged with the CCP programme mainly due to the financial and political resources it offered, and the legitimacy it provided for local climate protection actions (Betsill and Bulkeley 2004). In terms of resources, ICLEI and the CCP programme provides key resources, methods and tools for municipalities such as software for GHG inventories and preparing climate change action plans. Related ICLEI resources include reports, tools, case studies, briefing sheets, annual reports and climate action projects online (ICLEI 2010b; 2011). However, the reliability of CCP software for calculating urban and community GHG emissions was a key area of concern for LG, and greater input was needed in this area.

Organisational responsibility

Accountability for CCP rests with regional divisions or national offices of ICLEI and approved national partners, such as the Federation of Canadian Municipalities. LG input into CCP, within a country, appears to be limited. There is no external auditing of reports submitted to ICLEI by CCP members on their progress in achieving CCP milestones (Hoff 2010). With no enforcement or external monitoring system in place for CCP, many cities take minimal actions to reduce GHG emissions (Krause 2011: 45–60). No information is provided by ICLEI as to how CCP membership fees are utilised or how input and feedback is sought from LG. However, ICLEI is transparent in listing the CCP milestones and targets set by member municipalities. This CCP information is readily available on relevant ICLEI and LG member websites. Not all CCP members though provide publicly available reports on progress towards CCP targets.

Decision-making

It is unclear how ICLEI has involved CCP cities and climate action partners in democratic decision-making about the implementation of CCP. ICLEI USA involved its LG stakeholders in the steering committee and six technical advisory committees developing the new Community-Scale GHG Emissions Analysis Protocol (Ewing-Thiel and Manarolla 2011: 371–375). In Australia, some CCP partners found the national ICLEI office to be didactic in dealing with LG members. Local politics also limits the effectiveness of urban climate governance with varied spatial perspectives and time-scales for climate actions (Stoker 2000: 91–109). In terms of agreement, CCP municipalities pledge to reduce their emissions, select their milestone (Levels 1–5), prepare climate plans and set GHG reduction targets. However, many CCP cities only commit to lower levels of climate action (Gore 2010: 37–46), while GHG goals are voluntary and few climate actions may be adopted (Wheeler 2008: 481–491). There are no penalties for cities not meeting GHG reduction targets, no opportunity for CCP cities to challenge or verify GHG claims, and no formal mechanism is outlined for dispute settlement within CCP as a voluntary programme.

Implementation

In terms of behavioural change, CCP has been successful in changing the behaviour and organisation of LG participants in Australia and New Zealand to include climate change actions as part of municipal operations and allocating staff to this area (Hoff 2010). CCP is influencing climate change behaviours at the local level among mainly larger cities and metropolitan municipalities. The activities of climate protection networks such as CCP influence the adoption of local city policies to reduce GHG emissions (Bulkeley and Betsill 2005). However, cities that own electricity utilities or rely on manufacturing are less likely to commit to climate protection programmes such as CCP (Krause 2011: 45–60). With regard to problem solving, CCP assists LG to address climate change issues. This includes outlining LG strategies and actions to reduce GHG emissions. In terms of durability, CCP has been operating since 1993 as the main global LG climate network. The success of CCP has also influenced similar urban climate networks in the United States (for example Mayors Climate Protection Agreement) and Europe (that is, Covenant of Mayors). CCP may be more effective in supporting early commitment by LG to climate mitigation actions, with more experienced municipal members requiring a higher level of recognition to secure their ongoing investment and involvement in the CCP programme. In Australia, the number of CCP members decreased after federal government financial support for CCP ended in 2009. Globally, however, CCP has played a significant role in urban climate programmes, albeit with varied governance quality. This is evident in the tensions between CCP as a global programme requiring local responses.

Conclusions and recommendations

Municipal organisations like ICLEI are now key actors in multi-level global climate governance, with LG now formally recognised as a key governmental stakeholder at the COP 16 summit. Local authorities have less influence on national or sub-national climate change policies, but many still voluntarily adopt climate mitigation actions though CCP and related programmes. Governance issues for these transnational climate networks include the relationship between LG climate networks and participating cities, between municipalities in climate programmes and between national and international climate networks for cities. Large cities in the CCP programme are often policy leaders in climate action with growing influence in national and global climate governance regimes.

The governance quality of CCP varies according to whether it is assessed at a global or local level. The governance strengths of CCP were in interest representation and implementation of LG climate actions. However, at a global or national level, CCP lacked governance capacity in terms of organisational responsibility and decision-making to include input and feedback from LG. The governance of CCP by ICLEI, linked with other climate networks, has had a global impact in terms of LG inclusion in climate negotiations since 2010. However, LG participation in CCP is by the larger cities and towns and by metropolises in developed countries. The governance of urban climate networks such as ICLEI and the CCP programme is also driven by the global climate agenda dominated by developed countries.

The CCP programme has evolved from a civic environmental discourse with cities as stakeholders to a green governmentality, based on a scientific and business approach (Bäckstrand and Lövbrand 2007: 123–148). ICLEI and CCP now have a business-based approach to assessing urban climate actions through the International LG GHG Emissions Analysis Protocol (IEAP), the GPC-Scale Greenhouse Gas Emissions and Cities Climate Registry. Accountability has been improved through this shift, particularly in the standardisation of GHG reporting by cities – but it may also exclude smaller LG that are adopting a narrower range of climate mitigation initiatives at the local level. While organisations allied with ICLEI such as C40 Cities and the CDP do report the GHG emissions of large cities, most assessment of GHG emissions by LG is still voluntary.

Specific recommendations for ICLEI policy and governance of CCP include the following:

- LG involvement in assessing reports submitted by CCP members within each country;
- Involving LG members in ICLEI's collaboration with other municipal/urban climate networks; and
- Supporting a diverse range of CCP members (north/south; large/small) at global climate negotiations.

These measures will enhance the strategic role of LG networks in urban climate governance.

References

Anguelovski, I. and Carmin, J.A. 2011. Something borrowed, something new: innovation and institutionalization in urban climate governance. *Current opinion in environmental sustainability* 3 (3), 169–175.

Arikan, Y. 2011. *Carbonn Cities Climate Registry 2011 annual report: local response to measurable, reportable, verifiable global climate action.* Bonn Centre for Local Climate Action and Reporting, http://www.citiesclimateregistry.org/fileadmin/user_upload/cCCR/cCCR_2011_Annual_Report/cCCR_2011_Annual_Report_www_lo_20111125.pdf, accessed 19 June 2012.

Bäckstrand, K. and Lövbrand, E. 2007. Climate governance beyond 2012: Competing discourses of green governmentality, ecological modernization and civic environmentalism. *In:* M. Pettinger, ed. *The social construction of climate change: power, knowledge, norms, discourses.* Aldershot and Burlington, VT: Ashgate, 123–148.

Bassett, E. and Shandas, V. 2010. Innovation and climate action planning: perspectives from municipal plans. *Journal of the American planning association*, 76 (4), 463–475.

Betsill, M.M. and Bulkeley, H. 2007. Looking back and thinking ahead: a decade of cities and climate change research. *Local environment*, 12 (5), 447–456.

Betsill, M.M. and Bulkeley, H. 2006. Cities and the multilevel governance of global climate change. *Global governance*, 12 (2), 141–159.

Betsill, M.M. and Bulkeley, H. 2004. Transnational networks and global environmental governance: the Cities for Climate Protection program. *International studies quarterly*, 48 (2), 471–493.

Boswell, M.R., Greve, A.I. and Seale, T.L. 2010. An assessment of the link between greenhouse gas emissions inventories and climate action plans. *Journal of the American planning association*, 76 (4), 451–461.

Bouteligier, S. 2011. Inequality in new global governance arrangements: the North South divide in city networks for global environmental governance. *In: Inequality in global governance: causes and consequences of unequal representation and decision-making in global governance institutions.* Belgium: GRESI/FWO International Workshop, Leuven Centre for Global Governance Studies, 8–9 November, www//ghum.kuleuven.be/ggs/events/2011/11_2011/workshop-8-9nov2011.pdf, accessed 21 June 2012.

Bulkeley, H. and Betsill, M. 2005. Rethinking sustainable cities: multilevel governance and the 'urban' politics of climate change. *Environmental politics*, 14 (1), 42–63.

Bulkeley, H. and Betsill, M. 2003. *Cities and climate change: urban sustainability and global environmental governance*. London: Routledge Press.

Cadman, T. 2011. *Quality and legitimacy of global governance: case lessons from forestry*. Basingstoke and New York: Palgrave Macmillan.

Carbon Disclosure Project. 2011. *CDP cities 2011 global report on C40 Cities*. http://www.c40citieslive.squarespace.com/storage/resources/CDP%20Cities%202011%20Global%20Report.pdf, accessed 19 June 2012.

———. 2010. *Carbon disclosure project: the case for city disclosure*. Accenture. Retrieved 19 June 2012, from Carbon Disclosure Project Web site: http://www.cdproject.net.

Chavez, A. and Ramaswami, A. 2011. Progress toward low carbon cities: approaches for transboundary GHG emissions' footprinting. *Carbon management*, 2 (4), 471–482.

Coenen, F. and Menkveld, M. 2002. The role of local authorities in a transition towards a carbon-neutral society. *In:* M.T.J. Kok, W.J.V. Vermeulen, A.P.C. Faaij and D. de Jager, eds. *Global warming and social innovation: the challenge of a climate neutral society*. London: Earthscan Publications, 107–125.

Corfee-Morlot, J., Kamal-Chaoui, L., Donovan, M., Cochran, I., Robert, A. and Teasdale, J. 2009. *Cities, climate change and multilevel governance*, OECD Environmental Working Papers No. 14. Paris: OECD Publishing.

C40 Cities. 2011. *Climate action in megacities: C40 cities baseline and opportunities*, http://www.live.c40cities.org/storage/ARUP%20REPORT%20-%20Climate%20Action%20inb%20Megacities.pdf, accessed19 June 2012.

EU Mayors. 2012. Covenant of Mayors. *Signatories*. Retrieved 22 June 2012, from EU Mayors Web site: //http://www.eumayors.eu/about/signatories_en.html.

Ewing-Thiel, J. and Manarolla, X. 2011. Policy update: ICLEI USA draft framework for measuring and reporting GHG emissions. *Carbon management* 2 (4), 371–375.

Gore, C.D. 2010. The limits and opportunities of networks: municipalities and Canadian climate change policy. *Review of policy research*, 27 (1), 27–46.

Gore, C. and Robinson, P. 2009. Local government response to climate change: Our last, best hope? *In:* H. Selin and S. VanDeveer, eds. *Changing climates in north american politics: institutions, policymaking, and multilevel governance*. Cambridge MA: MIT Press, 137–158.

Granberg, M. and Elander, I. 2007. Local governance and climate change: reflections on the Swedish experience. *Local environment*, 12 (5), 537–548.

Hamin, E.M. and Gurran, N. 2009. Urban form and climate change: Balancing adaptation and mitigation in the U.S. and Australia. *Habitat international*, 33 (3), 238–245.

Hoff, J. 2010. *Local climate protection programmes in Australia and New Zealand – results, dilemmas and relevance for future actions*. CIDEA Project Report No. 1. Denmark: CIDEA.

International Council for Local Environmental Initiatives (ICLEI). 1997. *Local government implementation of climate protection: a report to the United Nations*. Toronto: ICLEI.

International Council for Local Environmental Initiatives (ICLEI). 2012a. *2012 ICLEI world congress program book*, http://www.iclei.org/worldcongress2012, accessed 19 June 2012.

———. 2012b. ICLEI at Bonn climate talks., http://www.iclei.org, accessed 19 June 2012.

———. 2011. *Global environmental governance: the role of local governments*, http://www.iclei.org/fileadmin/user_upload/documents/Global/Services/Resources/ICLEI_Paper_2011-1_IEG_.pdf, accessed 19 June 2012.

———. 2010a. *The city climate catalogue*, http://www.climate-catalogue.org/index.php?id=6905, accessed 19 June 2012.

———. 2010b. *Cities in a post-2012 climate policy framework*, http://www.iclei.org/fileadmin/user_upload/documents/Global/Services/Resources/Cities_in_a_Post-2012_Policy_Framework-Climate_Financing_for_City_Development_ICLEI_2010.pdf, accessed 19 June 2012.

———. 2006. *Combating climate change: a comprehensive look at local climate protection programs*, http://www.iclei.org/documents/USA/services/CaseStudiesDec20062.pdf, accessed 19 June 2012.

———. 2005. *World mayors and municipal leaders demonstrate leadership on climate change*, http://www.iclei.org/fileadmin/user_upload/documents/events/montreal_summit/Media_and_Reports/ICLEI_Montreal_Summit_PR_Dec_8.pdf, accessed 19 June 2012.

ICLEI Oceania. 2011. *Cities for climate protection*. Retrieved 19 June 2012, from ICLEI Web site: http://www.iclei.org/index.php?id=10681.

Kamal-Chaoui, L. and Robert, A. (eds.) 2009. *Competitive cities and climate change*. OECD Regional Development Working Papers No, 2. Paris: OECD Publishing, http://www.fatf-gafi.org/dataoecd/30/36/44232251.pdf, accessed 19 June 2012.

Krause, R.M. 2011. Policy innovation, intergovernmental relations, and the adoption of climate protection initiatives by U.S. cities. *Journal of urban affairs*, 33 (1), 45–60.

Metropolis. 2011. Cities and climate change. *In: Ecological region commission 1 report*, http://www.metropolis.org/sites/default/files/publications/2011/C1%20Metropolis%20Eco%20regions-english.pdf, accessed 19 June 2012.

———. 2009. *Climate change: cities in action*. Metropolis International Institute., http://www.metropolis.org/sites/default/files/publications/2009/Climate-Change-Cities-in-Action_0.pdf, accessed 19 June 2012.

Okereke, C., Bulkeley, H. and Schroeder, H. 2009. Conceptualizing climate governance beyond the international regime. *Global environmental politics*, 9 (1), 58–78.

Pearce, G. and Cooper, S. 2011. Sub-national responses to climate change in England: evidence from local area agreements. *Local government studies*, 37 (2), 199–217.

Rabe, B.G. 2007. Beyond Kyoto: climate change policy and multilevel governance. *Governance* 20 (3), 423–444.

Ross, A. 2011. *What future now for the wider sustainability agenda?* http://www.idea.gov.uk/idk/core/page.do?pageld+27962383, accessed 19 June 2012.

Satterthwaite, D. 2008. Cities' contribution to global warming: notes on the allocation of greenhouse gas emissions. *Environment and urbanization*, 20, 539–549.

Slocum, R. 2004. Consumer citizens and the Cities for Climate Protection campaign. *Environment and planning A*, 36 (5), 763–782.

Stoker, G. 2000. Urban political science and the challenge of urban governance. *In:* J. Pierre, ed. *Debating governance: authority, steering and democracy*. Oxford: Oxford University Press, 91–109.

Storey, H., Brennan, M., Pillora, S. and Thomas, C. 2012. *Local action for a low carbon Future*. Discussion Paper June 2012. Sydney: Australian Centre of Excellence for Local Government, University of Technology Sydney.

United Cities and Local Government. 2011. *Climate change*, http://www.cities-localgovernments.org/sections.asp, accessed 19 June 2012.

UN. 2009. *Local authorities' initiatives in support of Agenda 21*, http://www.un.org/esa/dsd/agenda21/res_agenda21_28.shtml, accessed 19 June 2012.

UNEP. 2012. *Cities and municipalities*. Climate Neutral Network, http://www.unep.org/CLIMATENEUTRAL/Default.aspx?tabid=349, accessed 19 June 2012.

Wheeler, S.M. 2008. State and municipal climate change plans. *Journal of the American planning association*, 74 (4), 481–491.

The World Bank. 2010. *Cities and climate change: an urgent agenda*. Urban Development Series Knowledge Papers. Vol. 10. Washington DC: The World Bank.

Yienger, J., Brown, L. and Skinner, N. 2002. Experiences of ICLEI's Cities for Climate Protection campaign (CCP): a focus on Asia. *In: Proceedings of IGES/APN Mega-City Project*, 23–25 January 2002. Institute for Global Environmental Strategies.

15
The Influence of Non-State Actors on Corporate Climate Change Disclosure

Julie Cotter

Introduction and background

Disclosure of information about corporate activities relevant to the climate change phenomenon has increased substantially over the past decade. Climate change disclosures include those about greenhouse gas (GHG) emissions intensity and energy use; participation in emissions trading schemes; corporate governance and strategy in relation to climate change; performance against GHG emissions reduction targets; and regulatory, physical and other risks and opportunities of climate change. While various regulatory requirements to disclose information about GHG emissions and risks related to climate change are emerging in some jurisdictions, climate change-related disclosure in annual and/or sustainability reports or on company websites remains largely voluntary.

This chapter demonstrates how non-governmental organisations (NGOs) have played a necessary and important role in driving the increase in corporate climate change disclosure, something that national regulators have been unable or unwilling to do. Some governments and/or their agencies have passed legislation (and related guidance) requiring the disclosure of some aspects of climate change information. However, regulation of climate change reporting varies substantially between countries and tends to be limited to a narrow set of disclosures focused on the measurement of emissions. Disclosure of other types of climate change information such as related risks and opportunities is less regulated. While some countries require or encourage listed companies to disclose information about non-financial performance and material risks and opportunities, these requirements tend not to be specific to climate change. The reporting void created by the lack of decisive and collaborative state action has been lessened considerably by the global activities of several NGOs, including the Global Reporting Initiative (GRI), the Carbon Disclosure Project (CDP) and the Climate Disclosure Standards Board (CDSB).

There are several potential reasons why states have not mandated comprehensive climate change disclosures by corporations operating within their jurisdictions. First, there remains considerable uncertainty around the timing and impacts of climate change for particular regions and organisations. Second, the information requirements of company stakeholders in relation to climate change also lack a high degree of certainty. It is not yet clear exactly what information the various groups of corporate stakeholders need, where and how the information should be reported and to what level it should be independently assured, although the activities of NGOs described in this chapter continue to play an important role in reducing these reporting uncertainties. Third, increased mandatory disclosure and its assurance impose additional compliance costs on reporting entities, and this is something that governments tend to be reluctant to do without a highly defensible imperative. Indeed, it appears that the evolution of corporate climate change disclosure has not yet reached a level of maturity substantiating widespread regulatory intervention. Further, failure of governments to reach international agreement on how to deal with climate change has only served to exacerbate the reluctance to deliver a global standard for corporate climate change reporting.

Much of the existing national legislation has been put in place to facilitate proposed or actual emissions trading schemes. An example of this is Australia's National Greenhouse and Energy Reporting. Several other jurisdictions have requirements similar to Australia. A different form of regulation exists in the United States, where the Securities and Exchange Commission (SEC) provides public companies with interpretive guidance on existing SEC disclosure requirements as they apply to business or legal developments relating to the issue of climate change. The rules cover a company's risk factors, business description, legal proceedings and management discussion and analysis (SEC 2010). This regulatory intervention arose following large investor groups and pension funds petitioning the SEC. In France and Sweden, legislation has been provided to support the European Union Accounts Modernisation Directive (EUAMD), which requires certain companies to include a balanced and comprehensive analysis of the development and performance of their business in the Director's Report, including both financial and, where appropriate, non-financial key performance indicators relevant to the business and information on environmental matters (EUAMD 2003). Across the European Union (EU), most Member States have introduced guidance for compliance with the EUAMD. More recently, in 2012 the European Commission's Environment Directorate has issued a consultation on proposed mandatory environmental reporting for large European companies.

Notwithstanding this limited and internationally incoherent regulation of corporate climate change disclosure, there is evidence of a growing demand for increased disclosure and for standardised reporting guidelines. This is at least in part due to a perception that external reporting of GHG emissions helps to enable emissions reductions and development of climate change strategies by driving companies to measure emissions (PWC 2010: 1). Similarly, requirements to disclose information about an organisation's climate change risks are potentially a way of ensuring that due consideration is given to the likelihood that climate change could pose significant physical, regulatory, litigation and reputational risks to corporations.

A potentially significant outcome from the UNFCCC 17th Conference of the Parties (COP 17, Durban) was an agreement for increased cooperation and collaboration between the United Nations Framework Convention on Climate Change (UNFCCC) and Caring for Climate, the United Nations' climate action platform for business. The meeting focused on ways to better align Caring for Climate with the UNFCCC and to provide a channel for climate leaders to contribute to the global climate policy agenda. Caring for Climate is endorsed by nearly 400 companies from 65 countries. Chief executive officers who support the Caring for Climate statement are prepared to set goals, develop and expand strategies and practices and to publicly disclose emissions as part of their existing disclosure commitment within the UN Global Compact framework (UN Global Compact 2011).

Various non-state actors have contributed to bridging the gap between government-backed regulations and stakeholder demand for corporate climate change disclosure by (a) developing disclosure guidance and/or (b) exerting pressure on emitting organisations to disclose comprehensive climate change information. These non-state actors include accounting standards boards, securities exchanges, accounting professional bodies, NGOs and NGO partnerships. NGOs have also been active in making calls to national governments and securities exchanges to mandate increased corporate climate change disclosure.

Corporate climate change disclosure and the role of non-state actors

NGOs can be defined as 'a non-profit association of citizens that operates independently of government or business structures and has non-commercial objectives related to environmental, consumer interest or sustainable development' (ISO TC 207 2003: 4). There has been a growth and increasing role for non-state actors in the global community. Courville discusses the increase in power of non-state actors and attributes it to a shift in economic power away from the national to international levels and the rise of globalisation, trade liberalisation and privatisation (2003: 271). Standard setting is a dimension of regulation that can be well served by non-state actors due to their ability to be flexible and sensitive to market innovations (Hutter 2006: 13). It is therefore not surprising that there has been a powerful and increasing role for NGOs in the corporate climate change disclosure arena.

Non-state actors working in the area of climate change disclosure include those operating in both national and international jurisdictions. The activities of accounting professional bodies, accounting standards boards and securities exchanges are generally restricted by national boundaries, although there are a growing number of international collaborations that assist with reducing these jurisdictional limitations. Their activities in relation to climate change disclosure have tended to be at the national rather than international level. Examples of voluntary disclosure guidance include the Accounting Standards Board's (UK) statement of best practice 'Reporting Statement: Operating and Financial Review (RS1)', which recommends that companies identify principal environmental risks in qualitative terms and the potential impact of those risks on their results; the Ontario Securities Commission Staff Notice 51–716 'Environmental Reporting'; the 'Recommendation on Climate Disclosure in Annual Filings' published by the Japanese Institute of Certified Public Accountants; and the Canadian Institute of Chartered Accountants 2008 'Building a better MD&A: Climate Change Disclosures'.

However, it has been NGOs rather than these traditional disclosure regulators that have been most influential in improving international climate change disclosure. NGOs have a comparative advantage to both national governments and national accounting and disclosure regulators when it comes to establishing global reporting frameworks. Further, they have taken the lead and pursued global climate change disclosure guidance in collaboration with accounting bodies such as the

International Accounting Standard Board and international accounting professional bodies such as the Association of Chartered Certified Accountants (ACCA).

This observation aligns with Dellas, Pattberg and Betsill's assertion that solutions to the challenges of global change are co-produced by a number of non-state actors whose authority is contested and whose legitimacy is often questionable (2011: 87). As Betsill, Pattberg and Dellas explain, new modes of agency arise from the dilemma that both the public sector (governments) and private sector (markets) alone fail to adequately address climate change and human welfare (2011: 3–4). In this situation, civil society actors are increasingly recognised as agents in their own right with significant authority including rule-making, where agents can be thought of as authoritative actors, and authority is understood as the ability to exercise power and legitimacy (Betsill et al. 2011: 87). Authority is achieved through expertise, knowledge and innovative capacity or simply by responding to the shortcomings of state actors. Increasingly in relation to corporate climate change disclosure, non-state actors work in tandem to achieve greater agency.

NGOs and NGO partnerships that have contributed include the World Resources Institute (WRI), World Business Council for Sustainable Development (WBCSD), Ceres, the GRI, the CDP, the CDSB and the Investor Network on Climate Risk (INCR). The WRI and WBCSD developed the GHG Protocol: A Corporate Accounting and Reporting Standard. This is the most widely used international accounting tool for government and business leaders to understand, quantify, and manage greenhouse gas emissions. Its initial core steering group comprised of members from environmental groups (such as the World Wide Fund for Nature (WWF), Pew Center on Global Climate Change, The Energy Research Institute) and from industry (such as Norsk Hydro, Tokyo Electric, Shell) to guide a multi-stakeholder standards development process.

GRI and CDP are the two predominant global organisations when it comes to providing performance indicators and schemes for corporate climate change reporting. The GRI is a network-based organisation that has developed the world's most widely used sustainability-reporting framework. The GRI's first reporting guidelines were released in 2000 and were revised in 2002 and 2006. They consist of reporting principles and guidance as well as a comprehensive set of standard disclosures. The standard disclosures include key performance indicators related to GHG emissions and energy use as well as a range of other environmental, social and governance indicators. Globally, over 1800 companies used GRI as a framework for sustainability reporting in 2010.

CDP sends out an annual request for climate change information, and in 2011 had over 3000 responses from large corporations across 60 countries. This has grown substantially from the 221 responses to the first information request sent to the Chairmen of FT500 Global Index companies in 2003. The information request takes the form of a comprehensive questionnaire covering a wide range of climate change topics of interest to investors. Information is made widely available through CDP. While some companies choose to provide a link to their CDP response on their corporate website, CDP reporting remains separate from mainstream corporate reports such as annual and sustainability reports.

CDP leverages market forces including shareholders, customers and governments to incentivise companies to measure and disclose their GHG emissions and climate change risk strategies. It has harnessed the stakeholder power of 655 institutional investors, holding $78 trillion in assets under management, and over 50 purchasing organisations. Indeed, the CDP has spearheaded collective action by institutional investors on climate change disclosure globally. It seeks greater transparency and action on climate change from companies and uses corporate engagement as its preferred form of interaction (Hebb 2008: 12).

CDP has also played a role in calling for national governments to support enhanced corporate climate change disclosure and to mandate these disclosures. For example, at COP 17 CDP called for 'integrated and harmonised reporting between countries, cities & companies: integrated and harmonised measurement, reporting and verification (MRV) between national governments (countries), sub-national governments (cities) and companies is incredibly important, as the information needs to be consistent and comparable'. Investor coalitions including the Principles for Responsible Investment (PRI), INCR and United Nations Environment Programme Finance Initiative (UNEPFI) have also made calls to governments to mandate corporate disclosure of material climate change risks. In addition, PRI's investor principles include a commitment to seek appropriate corporate disclosure on environmental, social and governance issues that include climate change.

The CDSB was formed by a group of NGOs including CDP to establish a comprehensive reporting framework for climate change information. It was formed at the 2007 annual meeting of the World Economic Forum (WEF) in response to increasing demands for standardised reporting guidelines on the inclusion of climate change information in mainstream reports. Board members include representatives from CDP (Secretariat), Ceres, The Climate Group, The Climate Registry,

International Emissions Trading Association, World Economic Forum, WBCSD and WRI.

CDSB has developed an international framework for corporate reporting on climate change. Its Climate Change Reporting Framework (CCRF) draws on the disclosure requirements of leading climate change disclosure initiatives, including The Climate Registry, GRI and CDP, and research conducted by the INCR coordinated by Ceres. These sources reveal general consensus that the types of information that should be considered in preparing climate change-related disclosures fall into four categories: strategy and governance, risks and opportunities, GHG emissions and activities a company takes to mitigate and/or adapt to climate change (CDSB 2009: 2). The CCRF adopts the GHG Protocol for preparing an emissions inventory, the de facto standard on emissions monitoring and measurement. Its aim is to link disclosures to information about financial performance and mainstream financial reports. To this end, the CCRF adopts relevant principles from financial reporting and seeks to establish the organisational boundaries used in financial reporting as the consolidation approach for climate disclosures (CDSB 2010: 5, 23). It elicits information that can be integrated into investor analysis and provides conceptual and practical input for regulatory agencies. CDSB is supported by the accounting profession (multinational accounting firms and professional bodies) through its technical working groups.

Overall, the role of non-state actors, particularly NGOs and NGO partnerships, on corporate climate change disclosure has been considerable. Their ability to work beyond national boundaries has assisted them to develop globally recognised reporting schemes and frameworks that allow multinational and other large companies to demonstrate their corporate citizenship in relation to climate change. These voluntary reporting frameworks are flexible enough to accommodate international differences, while at the same time encouraging reporting against a standard set of criteria. Investor coalition NGOs have also been active in making calls on governments to play a greater role in facilitating corporate climate change disclosure.

Critical analysis

As indicated earlier in this chapter, while various state and non-state actors have made substantial progress with developing guidance for corporate climate change disclosure, its evolution does not yet appear to have reached a level of maturity sufficient to substantiate widespread government-backed regulation. However, it is not clear whether the current governance arrangements, which rely to a large extent on the actions and integrity of non-state actors, are the most effective for the medium and longer term. Further, the lack of mandated, standardised reporting means that there are material uncertainties and weaknesses in the quality and extent of corporate climate change disclosure.

Weaknesses in corporate climate change disclosure

Adoption of voluntary CCRFs has been relatively limited, especially among smaller companies. CDP is the primary climate change disclosure vehicle for multinational and very large companies. The largest companies tend to report both via completion of the CDP questionnaire and in corporate reports prepared following the GRI guidelines. Cotter and Najah studied the climate change disclosures of a global sample of 340 large, non-financial firms. They found that these companies tend to disclose only a relatively small portion of the information that they provide to CDP in their annual and sustainability reports. However, even for those large corporations that make climate change disclosures, there remains substantial variation in the amount, type and quality of information (2012: 14).

Corporate climate change disclosure weaknesses relate to comprehensiveness, quality and comparability of the information provided. The CDP and CDSB schemes are comprehensive in terms of the type and extent of disclosures required, and it could be argued that CDSB's framework is a more standardised and decision-useful version of CDP's questionnaire. However, GRI is less comprehensive when it comes to climate change information. Disclosure requirements of governments and their agencies tend to be even less comprehensive and focus primarily on measurement of emissions to support emissions trading schemes. A summary of the specific disclosure content areas required by each of the major reporting schemes is presented in Table 15.1.

Cotter and Najah rated the comprehensiveness and quality of climate change disclosures for their sample companies out of a possible score of 100. The average climate change disclosure score for these companies is 19, with a maximum of 51. 'Emissions accounting' and 'governance' have the highest scores, while 'risks and opportunities' is the category with the lowest scores. Indeed, 38 per cent of their sample of large companies did not disclose anything about climate change risks or opportunities. Materials and utilities sector companies disclosed the most, while those from the industrial, energy, telecommunication and health-care sectors were the worst disclosers (2012: 12–13).

Table 15.1 Specific disclosure requirements by reporting framework

Disclosure requirement	**GRI**	**CDP**	**CDSB**
GHG emissions accounting	✓	✓	✓
Energy use	✓	✓	✓
Strategy and governance	✓	✓	✓
Risks and opportunities	–	✓	✓
Mitigation and adaptation activities	–	✓	✓
Standardised corporate reporting framework	✓	–	✓
Reporting principles	✓	–	✓

Cotter, Najah and Wang provide an in-depth analysis of the climate change disclosures for a company that has received awards for its climate disclosure record. They conclude that the disclosures tend to lack technical detail and are somewhat skewed towards the more positive aspects of climate change impacts and management (2011: 17). Similar criticisms have been made about the quality of corporate sustainability reports prepared in accordance with the GRI guidelines. Poor climate change disclosure quality is due at least in part to the limited verification or assurance generally associated with voluntary reporting.

Criticisms of CDP questionnaire responses relate to lack of standardisation and quality of disclosures. Kiernan contends that even experienced analysts of climate change and emissions data find it very hard to make sense of firm reporting as part of CDP (2008: cited in Kolk et al. 2008: 741). Based on an examination of corporate responses to climate change in relation to the development of carbon disclosure mechanisms, Kolk et al. conclude that voluntary carbon disclosure remains 'inconsistent and difficult to interpret' (2008: 721). Investors are grappling with how to use the information that is being made available or are choosing not to use it. Similarly, there is an expressed demand for standardised reporting guidelines to facilitate comparability between companies and improve the quality of climate change disclosures. Smith, Morreale and Mariani conclude that what is required is the development of an effective language for climate disclosure embodied in a detailed set of guidelines that companies can then be required to apply to their own, widely divergent business and prospects, for all material matters (2008: 485).

The CDSB draft reporting framework provides a first step in this direction. It offers a conceptual and practical input for regulatory agencies by working to develop a comprehensive international framework for corporate reporting on climate change. An aspect of this framework that is particularly appealing is that it aligns with the International Accounting Standards Board's Conceptual Framework and notions of decision-usefulness and other qualitative characteristics (CDSB 2010: 14). However, the reporting rigour and comprehensiveness required may act as a deterrent to companies contemplating increased climate change disclosure and to national governments considering compliance costs related to mandatory climate change reporting requirements.

Assessment of non-state actor governance of climate change disclosure

As demonstrated in this chapter, the governance arrangements for corporate climate change disclosure rely to a large extent on the actions and integrity of non-state actors. These organisations function, for the most part, outside the traditional nationally based political establishment (Bulkeley 2010: 312). In contrast to global NGO initiatives to enhance and standardise corporate climate change disclosure, there has been a piecemeal emergence of divergent national requirements that are limited in scope and often exist outside of mainstream corporate reporting. Placing considerable reliance on non-state actors raises questions about the legitimacy and effectiveness of the current governance arrangements.

Other questions relate to what is needed for national governments and their agencies to play a greater role, and how any shortcomings can be addressed. As previously indicated, it appears that the evolution of corporate climate change disclosure has not yet reached a level of maturity substantiating widespread regulatory intervention. This is at least partly due to concerns associated with the cost of imposing mandatory disclosure requirements and uncertainty around climate change impacts and stakeholder information needs. Further, the legitimacy of prescribed reporting frameworks and the organisations involved in their development is extremely important for regulators considering mandatory corporate climate change disclosure.

Institutional investors represent a powerful and legitimate stakeholder group for companies in which they hold substantial shareholdings. They are also a group that has expressed a desire for high-quality information about corporations' exposure to risks associated with climate change (Stanny and Ely 2008: 338). Indeed, Smith, Morreale and Mariani conclude that the response of institutional investors to voluntary, market-based disclosure initiatives indicates that investor demands for climate disclosure have driven private action faster than either regulators or politicians have addressed the underlying issues (2008: 474). The activities of CDP demonstrate how the NGO sector can essentially capture one or more major stakeholder groups to exert pressure on emitting organisations. The research confirms that secondary stakeholders, such as NGOs, are key players in CSR, but their role is still regarded as controversial and their legitimacy contested (Arenas et al. 2009: 183).

Legitimacy of NGO governance is an ongoing problem since NGOs are not accountable to the states they represent, if not the general public directly. Internally, the decision-making processes of many NGOs are not exposed to members, procedures for electing their leaders are absent or they may not even have members (Rosenau 2000: 193). Further, NGOs are not always transparent (Keohane 2003: 144–145). Cadman's normative governance framework (Introduction, this volume) provides a benchmark for evaluating the legitimacy of non-state actors involved in corporate climate change disclosure. Using this governance assessment tool, GRI and CDSB would arguably rate higher than CDP in terms of interest representation, organisational responsibility and democratic decision-making processes. Although CDP acts as Secretariat to CDSB, and has therefore played a major role in creating the greater legitimacy that CDSB can claim, GRI prides itself on being a multi-stakeholder, network-based organisation that strives to provide meaningful participation for its many stakeholders. Its international working groups, stakeholder engagement and due process further support the legitimacy of GRI and its reporting guidelines. Similarly, CDSB facilitates participation and decision-making through its advisory committee, technical working group and engagement with investors, preparers, governments and other relevant stakeholders.

The disclosure schemes and frameworks produced by each of these organisations can be assessed in terms of their effectiveness, including the extent to which the frameworks have been adopted, and corporate behavioural changes leading to increased accountability and transparency on climate change. In addition, they can be assessed by their contributions towards solving the underlying problems of climate change and their durability.

The results of Cotter and Najah's research provide support for the contention that the influence of CDP is positively associated with climate change disclosure by large companies. They find the extent and quality of climate change disclosures to be associated with three indicators of corporate responsiveness to CDP's annual requests for climate change information. These are completion and publication of the CDP questionnaire on CDP's website, indications in corporate communications that CDP activities have influenced climate change disclosures, and the extent and quality of climate change information provided in CDP questionnaire responses. They find that 49.5 per cent of Global 500 companies completing the CDP questionnaire made some sort of reference to CDP in their annual and/or sustainability report disclosures, thus providing an explicit indication that CDP activities influenced their climate change disclosure choices (2012: 11).

The GRI reporting guidelines also appear to have increased corporate accountability and transparency in relation to climate change. GRI has improved the comparability and quality of sustainability reporting, including climate change-related KPIs, for a considerable number of large organisations globally. On the other hand, the CCRF developed by CDSB is relatively new and there is as yet limited evidence on the extent of its adoption. Notwithstanding the criticisms of government-mandated climate change disclosure requirements discussed above, adoption of these schemes tends to be high relative to voluntary schemes.

There are also likely to be benefits such as improved emissions management associated with the increased disclosures achieved by the various voluntary and mandatory reporting schemes. However, it is as yet unclear to what extent increased corporate disclosure has been the catalyst for carbon reductions. Further research into this important issue is needed. Similarly, Dellas, Pattberg and Betsill call for more research to address a gap in the research literature in relation to evaluation of the effectiveness of agents such as the non-state actors considered in this chapter (2011: 96).

The durability of the various schemes will depend on their ongoing relevance in a dynamic reporting environment. If widespread adoption of voluntary reporting schemes such as the CCRF is achieved, or if their application receives regulatory support in a large number of countries, the CDP may lose its relevance. However, until such change is achieved, the CDP is expected to continue to play a major role in increasing the flow of corporate information about climate change.

The corporate reporting environment continues to evolve and there are moves towards greater integration and harmonisation, across both national boundaries and the various reporting frameworks. The International Integrated Reporting Committee (IIRC) has been established with a mission to create a globally accepted integrated reporting framework that brings together financial, environmental, social and governance information in a clear, concise, consistent and comparable format. It is unlikely that climate change reporting will remain as a stand-alone concept as more companies and countries move towards integrated reporting. However, it is likely that the IIRC will rely on CDP, CDSB and GRI for climate change content and that it will take many years to achieve widespread adoption of integrated reporting.

A recent initiative has been the formation of an inter-agency working group comprising the United Nations Conference on Trade and Development (UNCTAD), Organisation for Economic Cooperation and Development (OECD), CDSB and GRI to assess the benefits of greater consistency of approach to corporate climate change reporting. This will pave the way for further action to support consistency in the demand for and supply of climate change-related information. In support of this initiative, the OECD is undertaking a comparative case study analysis of voluntary and mandatory carbon reporting provisions in the United Kingdom, France, Australia and Japan. Preliminary findings indicate that divergence in climate change reporting schemes are due to different motivations for governments to put in place carbon reporting provisions, the ultimate use of information and different target audiences. Alignment with legitimate and visible international organisations such as the United Nations and OECD further enhances the legitimacy of CDSB and GRI, and this has the potential to make the outcomes of this initiative attractive to national governments and their agencies grappling with the issue of when and how to increase regulatory support for corporate climate change disclosure.

Independent research evidence on corporate stakeholder climate change information needs would also assist national disclosure regulators to assess the extent and nature of climate change disclosures to mandate. While research conducted or supported by non-state actors such as CDSB provides valuable input to regulatory decision-making, its lack of independence may serve to reduce its persuasiveness.

Conclusions and recommendations

The internationally recognised tool for measuring GHGs, the GHG Protocol, was developed by an NGO partnership (WRI and WBCSD) and forms the basis of the 14064 ISO standards. Two NGOs, the GRI and the CDP, have been particularly influential in terms of their impact on corporate reporting. Many multinational and other large companies have embraced climate change reporting and sustainability reporting more broadly as a way of demonstrating their corporate citizenship. A large proportion of these use the GRI and/or CDP frameworks to guide this reporting. In responding to these demands, the CDSB has developed a comprehensive and more standardised international framework for corporate climate change reporting: the CCRF.

GRI, CDP and CDSB have different strengths when evaluated against Cadman's normative governance framework. GRI and CDSB rate well in terms of interest representation, organisational responsibility and democratic decision-making, with GRI being particularly strong in these areas. However, CDP appears to have been the most effective in increasing corporate climate change disclosure, most probably due to the legitimacy and power of the institutional investors it represents. The durability of the various schemes is uncertain and will depend to some extent on progress with and take up of integrated reporting. The recently formed inter-agency working group has the potential to further enhance the legitimacy of the GRI and CDSB voluntary reporting frameworks and to define a set of reporting guidelines that would be acceptable to a wide range of national disclosure regulators.

Possible explanations why national regulators have been reluctant to mandate comprehensive climate change disclosures include uncertainty around the timing and impacts of climate change for particular regions and organisations. A further source of uncertainty relates to the information requirements of company stakeholders in relation to climate change, while compliance costs are another potential deterrent.

CDP and CDSB explain that mandatory corporate climate change reporting has the potential to provide transparency and vital information for shareholders, as well as encourage behaviour change and carbon reduction among business. It also supports government objectives to manage and mitigate climate change (2012: 3).

If mandating additional corporate disclosures is still some time away for the reasons articulated above, alternative solutions could be considered in the interim. For example, it is expected that authoritative guidance rather than mandated climate change disclosure requirements will be preferred in many countries. It is likely that authoritative guidance (at least) will be needed to achieve climate change reporting on a global scale. However, only a rigorous reporting framework will provide the consistency and quality demanded by investors.

Governance and policy recommendations

Based on the discussion presented in this chapter, the following courses of action are suggested:

- NGOs should continue working to extend their corporate climate change disclosure reach and influence beyond very large companies.
- NGOs should identify and address the reasons why national regulators have been reluctant to mandate comprehensive climate change disclosures.
- National regulators should mandate a comprehensive set of climate change disclosure requirements, preferably based on a rigorously determined disclosure framework. A possible interim measure is government-sponsored authoritative guidance. This process would necessarily involve a consideration of the trade-offs between investor information demands, costs of compliance and a desire for concise reporting.
- Independent research evidence on corporate stakeholder climate change information needs would also assist national disclosure regulators to assess the extent and nature of climate change disclosures to mandate.
- Further research is needed to examine the validity of the perception that increased disclosure requirements would assist with driving emissions reductions and ensuring adequate consideration of climate change risks.

References

Arenas, D., Lozano, J.M. and Albareda, L. 2009. The role of NGOs in CSR: mutual perceptions among stakeholders, *Journal of business ethics*, 88 (1), 175–197.

Betsill, M., Pattberg, P. and Dellas, E. 2011. Editorial: agency in Earth Systems Governance. *International environmental agreements: politics, law and economics*, 11 (1), 1–6.

Bulkeley, H. 2010. Climate policy and governance: an editorial essay. *Wiley interdisciplinary reviews: climate change*, 1 (3), 311–313.

CDSB. 2010. *Climate change reporting framework* – edition 1.0, September. London: CSDB: http://www.cdsb.net/file/12/cdsb_copenhagen_update.pdf, accessed 5 December 2012.

CDSB. 2009. Climate Disclosure Standards Board Copenhagen update, COP 15, December: http://www.cdsb.net/file/72/climate-change-reportingframework-edition-1.0-jan-2012.pdf, accessed 5 December 2012.

Cotter, J., & Najah, M.M. 2012. Institutional investor influence on global climate change disclosure practices. *Australian journal of management*, 37 (2), 169–187.

Cotter, J., Najah, M. and Wang, S. 2011. Standardized reporting guidelines for climate change related information: an Australian case study. *Sustainability accounting, management and policy journal*, 2 (2), 294–321.

Courville, S. 2003. Understanding NGO-based social and environmental regulatory systems: why we need new models of accountability. *In*: M. Dowdle, ed.

Public accountability: designs, dilemmas and experiences. Cambridge: Cambridge University Press, 271–300.

Dellas, E., Pattberg, P. and Betsill, M. 2011. Agency in earth system governance: refining a research agenda. *International environmental agreements: politics, law and economics*, 11 (1), 85–98.

European Union Accounts Modernisation Directive (EUAMD). 2003. The modernisation of accounting directive 2003/51/EC.

Hebb, T. 2008. No small change: Pension funds and corporate engagement. *Digital commons*, Cornell University Press, http://digitalcommons.ilr.cornell.edu/books/42, accessed 1 June 2012.

Hutter, B. 2006. *The rise of non-state actors in regulation*, discussion paper 37, Centre for Analysis of Risk and Regulation, London School of Economics and Political Science.

ISO TC 207. 2003. *Increasing the Effectiveness of NGO Participation in ISO TC 207*. No location: ISO TC 207. http://www.pacinst.org/inni/NGOParticipation/N28Final.pdf, accessed 1 June 2012.

Keohane, R.O. 2003. Global governance and democratic accountability. *In*: D. Held and M. Koenig-Archibugi, ed. *Taming globalization: frontiers of governance*. Cambridge: Polity Press, 130–159.

Kiernan, M.J. 2008. Climate change and investment risk, Presentation at the Amsterdam Global Conference on Sustainability and Transparency, GRI, Amsterdam.

Kolk, A., Levy, D. and Pinske, J. 2008. Corporate responses in an emerging climate regime: The institutionalization and commensuration of carbon disclosure. *European accounting review*, 17 (4), 719–745.

Price Waterhouse Coopers. 2012. *Carbon disclosure project 2010 global 500 report*. No location: Carbon Disclosure Project: https://www.cdproject.net/CDPResults/CDP-2010-G500.pdf, accessed 5 December 2012.

Price Waterhouse Coopers and CDP Report. 2010. *Review of the contribution of reporting to GHG emissions reductions and associated costs and benefits*.

Rosenau, J. 2000. Change, complexity and governance in a globalising space. *In:* J. Pierre, ed. *Debating governance: authority, steering and democracy*. Oxford and New York: Oxford University Press, 167–200.

SEC. 2010. Commission guidance regarding disclosure related to climate change, 17 CFR PARTS 211, 231 and 241 [Release Nos. 33-9106; 34-61469; FR-82].

Smith, J.A., Morreale, M. and Miriani, M.E. 2008. Climate change disclosure: Moving towards a brave new world. *Capital markets law journal*, 3 (4), 469–485.

Stanny, E. and Ely, K. 2008. Corporate environmental disclosures about the effects of climate change. *Corporate social responsibility and environmental management*, 15 (6), 338–48.

UN Global Compact. 2011. Caring for climate signatories meet with UNFCCC secretariat, http://unglobalcompact.org/news/177-12-06-2011, accessed 7 May 2012.

Conclusion

All Contributors

Given the complexity of the actors and institutions involved in the climate regime complex, greater understanding of the relations between state and non-state participants is required. Evaluating the success of this regime complex is therefore also about looking at the social processes that drive climate change-related policymaking. Decisions need to be equitable and address the wide-ranging concerns held both by those communities most at risk from climate change and broader society. The rise of a whole series of new policy instruments points to the regulatory and implementation deficits permeating existing multilateral regimes. But the legitimacy of both new and old institutional approaches – in so far as they deliver long-term solutions that reduce the threat of climate change – is as yet unknown. Given the billions of dollars that will begin to flow into financing climate change response mechanisms, the choice confronting the global community is whether it is more appropriate to redesign governance frameworks by improving existing arrangements or create yet more new institutions.

Whatever the answer to this question, most present measures are exclusionary and institutionally distant from the public at large and those who will most bear the cost of predicted impacts. A central theme common to the chapters in this volume is that greater levels of collaboration between all actors in the policy discourses of climate change are needed. It is in the structural and procedural contexts of participation and deliberation that an assessment of the institutions charged with governing climate change management must be conducted. Scholarly accounts have detailed how non-state actors are able to influence intergovernmental negotiations. But it is also important to distinguish between 'significant' and 'non-significant' influence. Acknowledging a degree of inclusion and some influence is not the same as saying such input has equal standing to other more powerful interests. Both inclusiveness and equality are essential to effective interest representation. The climate negotiations under the United Nations Framework Convention on Climate Change (UNFCCC) and the Conferences of the Parties (COPs) are marked by poor interest representation for non-state actors (with the possible exception of business interests). Until the structural and procedural constraints confronting non-state actors involved in global negotiations are fully appreciated, it appears unlikely that they will, in fact, ever influence international negotiations in ways that matter.

This is reflected in the debate around mitigating or adapting to climate change. In terms of government-led responses to climate change, it appears that liberal-democratic states have struggled to balance the demands of those who want the major policy focus to be on mitigation of the enhanced greenhouse effect, with those advocating adaptation. Unless the impacts of climate change are of a scale to accelerate the rate of policy innovation, it looks like the balance of policy is likely to shift to adaptation. This represents a highly pragmatic policy development that seeks to accommodate the demands of those who are seeking ways to continue industrial activity and those who are sceptical about the threats posed by climate change. It marks a move away from the recognition of the need to take radical action to combat climate change.

If civil society and other less powerful interests want to have greater influence over policymaking, it is clear they should continue to push for greater substantive representation in intergovernmental climate change-related policymaking arenas. This is important in helping to prevent the promotion of solutions that may have adverse environmental and social consequences. But it is equally clear that they should not merely concentrate their activities in such venues, which can only ever partially accommodate the views of non-state interests. Active participation is essential on all institutional fronts. But it should also be remembered that non-governmental initiatives seeking to solve complex sustainability problems are not excluded from the same risks that currently attend governmental arrangements. If both types of negotiating forum fail to secure representation from economic, social and environmental constituencies, then their deliberations will likely be partial and the solutions that eventuate unworkable. It is critical that all institutions seeking to combat climate change focus on acknowledging the expertise of participants rather than their status or power. International, national and local responses require the collective knowledge of all interests threatened by climate change, and these interests require institutions that can effectively manage stakeholder input. Once invited into such venues, all participants must be genuine partners, not just targets for manipulation.

There is some evidence that suggests that there has been a degree of improvement in the extent of stakeholder participation in climate governance in recent years. This has partly been a consequence of the blending of state and non-state interests within the policy discourse of market-based solutions to climate change. To date, Reducing Emissions from Deforestation and Forest Degradation (REDD+) has shown itself to be relatively inclusive of environmental and social stakeholders as well as developing countries. REDD+ demonstrates the developments in participatory governance at the global level that have occurred since the United Nations Conference on Environment and Development (UNCED) and over the last 20 years. But concerns have been expressed about the extent to which such systems have the necessary capacity for some stakeholders to participate meaningfully. It is a constant reality that few participants in such programmes have the financial, institutional or technical resources to participate at the level required. On the deliberative front, concerns have also been expressed about the ability of REDD+ (and other market systems) to settle disputes effectively. This is a problem in view of the ongoing uncertainties over who holds tenure and rights to carbon offsets. REDD+ has yet to be fully implemented, so it is unknown if it will substantively contribute to bridging the North–South gap, identified as an essential component of the sustainable development agenda, or if it will remain a merely anecdotal-level response, beneficial to the few. This has proven to be the case with the Clean Development Mechanism (CDM), which has been taken up by those increasingly able to help themselves (China, India) at the expense of other developing countries. The CDM may also be doing little to offset the accelerating growth in emissions in adopting countries. The scale of the financing needed for the carbon policy infrastructure to operate as designed has not been forthcoming. Mechanisms used to meet the greenhouse gas (GHG) emissions targets of Kyoto have absented one of the world's worst greenhouse polluters, notably the United States. Moreover, the planned volumes of the funds flowing to the principal Kyoto policy measures – REDD+, the CDM and JI – have not materialised. These less than hoped for financial outcomes are due in part to the unattractiveness of the various Kyoto financing vehicles to fiduciary investors, notably a distinct lack of clarity as to how the World Bank is to work with private financiers, insurance companies and pension- and mutual funds.

The efforts of other regimes to develop effective policy responses to climate change have been similarly mixed. With regard to human health, climate change has irrevocably altered the policy environment, but not necessarily for the better. The complexity of impacts and of responses to climate change means that an already complex and diffuse system of global health governance will become even more difficult to construct and manage. For much of the world's population, the impacts of climate change will be felt most directly through changes to freshwater resources and ecosystems. It is little appreciated that many policy responses to climate change are having perverse impacts on water resources and freshwater ecosystems. Many mutually beneficial policies for conservation of water and reduction of GHG emissions have not been prioritised nationally or internationally. There are examples of good integrative practices at the international and national scales but many opportunities to exploit positive synergies and accelerate environmental reforms are being missed. There has been little interplay of the UNFCCC with other, water-related environmental agreements, and many climate response policies are destructive for freshwater ecosystems and resources. Governance reform is needed to better integrate climate change responses with sustainable water management. In terms of food availability, there have been some positive steps in the right direction. It is to be hoped that recent efforts towards collaborative coordination and policy convergence, on the basis of the right to food principles, and recognising climate change issues, will result in better quality of governance in this field. A potentially fruitful dialogue between food security and food sovereignty has been created. Looking at energy management, few people would dispute the need for systemic governance reform to include environmental, social, cultural, economic, technological and political variables. Along with the expectation of energy access for all come the imperatives to protect human health and the environment and a rapid transition away from 'dirty' energy sources towards the sustainable development and use of alternative sources.

However, some of the stakeholders worst affected by climate change have little say in any institutional venues. For climate change displacees, the world has yet to acknowledge that migration is a legitimate form of adaptation to changing circumstances and therefore eligible for policy action and associated funding. According to current arrangements under the Climate Fund (established to facilitate implementation of the financial aspects of intergovernmental agreements), adaptation projects are generally conceived of as structural or technological changes that are implemented at the national level. However, individuals will not want to wait and see what kind of adaptation plans their governments implement. If they have already experienced difficulties they may prefer to leave. It is important to remember, should the numbers of climate displacees escalate, that their ability to do so is guaranteed under the UN Declaration of Human Rights. For those seeking to advocate the rights of 'climate change displacees', there may be something to be gained by learning from the way in which 'gender' has been used in the climate change policy debate. In this instance non-state actors have strategically exploited the ideological and rhetorical components associated with both gender and climate change to influence policymaking processes. Here, the conceptual currency associated with gender, and, in particular, with women's marginalisation from resources, has been taken up by actors as a strategic mechanism to influence policy. This strategic use of conceptual currency to influence policy deliberations is applicable beyond this specific case. Marginalisation and displacement are both rhetorical concepts that encapsulate discourses of disempowerment and inequality. Other non-state actors have also been successful in capturing the climate change policy debate and transforming it into concrete action. Local and municipal governments have been particularly effective in this regard. Their activities represent an important reconfiguration of climate change management, taking it out of the 'macro' intergovernmental sphere and situating it within the 'meso' domain of the nation state, but actually implementing it at the 'micro', or urban, level of governance. Local governments demonstrate a high level of convergence and collaboration within climate networks and have put targeted and effective climate change programmes and policies in place. Municipalities clearly have an important role to play in climate change decision-making, post-Kyoto agreements and GHG reduction goals. Ironically, however, the influence of local government on climate change appears to be stronger at the global level than in persuading governments to take strong action at the national level. Networks are also challenged internally, however, by inequality of interest representation and continue to be influenced by larger, often developed country municipalities, so further reform is necessary.

Clearly, non-state actors play a vital role in filling gaps not currently addressed by national regulation. In the case of carbon disclosure, for example, demands for increased and more standardised corporate climate change information, including mandatory reporting requirements, have been driven by institutional investor coalitions, who have gone to the non-governmental sector to meet their needs. This is a positive development. But these systems are largely voluntary and there is substantial variation in the amount, type and quality of reporting. The lesson to be learnt here is that state or non-state, public or private, mandatory or voluntary, market-driven initiatives need to strive for continual improvement in their standards of governance, including interest representation, organisational responsibility, decision-making and implementation.

But whatever the specific challenges confronting humanity as it grapples with the consequences of climate change, governments must bear the burden of responsibility. Greater political will is needed to better integrate those governance systems that make decisions about climate change policy. Fragmentation is currently not working to the advantage of the climate regime complex or the other regimes that are affected by it. Throughout the existing UN Conventions, provision of additional funding dedicated to cross-sectoral implementation at the national and international levels would be a powerful, short-term incentive for better governance across the climate regime complex. Governments must also have the courage to take on powerful vested interests. It is both possible and feasible to confront powerful lobby groups whose products and actions are detrimental to the health and well-being of the general public, and there is no reason in principle or practice why states should not exercise their constitutional powers to regulate those industries responsible for GHG emissions for the public and environmental good.

If they wish to be seen as legitimate systems within the climate policy regime complex, there are some important considerations that should inform institutions when shaping their responses to climate change into the future:

- Recognition must be given to the historical antecedents of climate change, including power, interest and differential responsibility, and that these frame deliberations on future actions;
- The differential interests of various stakeholders and rights-holders need to be recognised and their perspectives accorded appropriate status in any negotiations;
- By implication, those directly and otherwise impacted by climate change must be included as key participants from the beginning of deliberations concerned with response strategies;
- It is essential that procedures designed for consultation and negotiation be planned and formulated jointly by all interests;
- Any decisions regarding land allocation must be fully considered in conjunction with all parties and taking all relevant economic, cultural, social and psychological factors into consideration;

- Some element of compensation and restitution for those nations, groups and individuals impacted by climate change (but who contribute little to such change themselves) should be a paramount consideration in discussions;
- All decisions should remain mindful of the principles of social justice and human rights, in accordance with the relevant laws and protocols;
- State and non-state actors should work collaboratively towards increased international consistency across mandatory and voluntary governance systems. This should be achieved through governance standards that are acceptable to international and national regulators – and are sufficiently flexible to accommodate the differences between all parties in those systems.

Finally, all those working to combat human-induced climate change need to accept that business-as-usual is not an option. The scientific consensus on the need for substantive change in a vast array of human systems is overwhelming.

Index

Note: the letters *f* and *t* following locators refer to figures and tables.

Printed and bound in Great Britain by
CPI Antony Rowe, Chippenham and Eastbourne